Geochemistry

Scrivener Publishing
100 Cummings Center, Suite 541J
Beverly, MA 01915-6106

Publishers at Scrivener
Martin Scrivener (martin@scrivenerpublishing.com)
Phillip Carmical (pcarmical@scrivenerpublishing.com)

Geochemistry

Concepts and Applications

Edited by

Inamuddin, Mohd Imran Ahamed, Rajender Boddula and Tariq Altalhi

Scrivener Publishing

WILEY

This edition first published 2021 by John Wiley & Sons, Inc., 111 River Street, Hoboken, NJ 07030, USA and Scrivener Publishing LLC, 100 Cummings Center, Suite 541J, Beverly, MA 01915, USA

For more information about Scrivener publications please visit www.scrivenerpublishing.com.

Wiley Global Headquarters
111 River Street, Hoboken, NJ 07030, USA

For details of our global editorial offices, customer services, and more information about Wiley products visit us at www.wiley.com.

Library of Congress Cataloging-in-Publication Data

ISBN 978-1-119-70997-8

Cover image: Pixabay.Com
Cover design by Russell Richardson

Set in size of 11pt and Minion Pro by Manila Typesetting Company, Makati, Philippines

FSC www.fsc.org
MIX
Paper from responsible sources
FSC® C013604

Contents

Preface

Geochemistry is a branch of earth science. Since it is a field of study that uses the tools and principles of chemistry to explain the mechanisms in geologic environments, it often focuses on determining processes that control the abundance and composition of minerals and their distribution in the earth's crust. Geochemistry also plays a vital role in environmental soil and water systems in identifying and modulating environmental problems, and in studying the composition, structure and processes of the earth. Therefore, this book helps in understanding the chemical composition of the earth and its applications. Since it explores the basic principles, concepts and applications of geochemistry and discusses its beneficial effects, bottlenecks, solutions, and future directions, it will be a very helpful reference source for undergraduate and postgraduate students, geochemists, environmentalists, chemists, researchers, engineers, R&D professionals, and faculties who are working in the fields of earth science, environmental chemistry and industrial technologies among others. In addition to case studies, topics such as chemical weathering, impacts on living beings and water, geochemical cycles, oxidation and redox reactions in geochemistry, isotopes, analytical techniques, medicinal, inorganic, marine, atmospheric, and environmental applications are presented. The information contained in the chapters summarized below was written by experts in the field from around the world, making it a unique archival reference guide.

Chapter 1 discusses the nature and occurrence of toxic geogenic contaminants in serpentinite geological systems. Human exposure pathways in occupational and non-occupational settings are then summarized and their human health risks are also presented. Finally, mitigation measures to reduce human health risks are discussed and future research directions are proposed.

Chapter 2 reports detailed information on the application of essential minerals for the treatment of human diseases, including dental problems, cancer, hypertension, cardiovascular diseases, and goiters. Special attention is also paid to the application of essential elements for the management of animal and plant diseases.

Chapter 3 highlights the benefits of the application of geochemistry in livestock, especially in the adequate maintenance of animal health and animal nutrition. The significance of numerous minerals that play several roles in the effective production of livestock is especially highlighted.

Chapter 4 provides detailed information on the significance of geochemistry in the identification of useful mineral elements which could boost agricultural production and their significant role in the decontamination of heavily polluted soil and water, as well as their benefits especially in the area of sustainable agriculture.

Chapter 5 details the cases of heavy metal contamination in soil using the ecological risk index. The major focus is the geochemistry and environmental impacts of arsenic, cadmium, chromium, copper, lead, nickel and zinc. The transfer of contaminants from sources to the higher trophic levels of the food chain is also summarized.

Chapter 6 discusses the geochemical applications of both traditional and non-traditional stable isotopes. Beginning with basic definitions, the chapter goes on to discuss the variety of applications of traditional stable isotopes, geothermometers, tracers in hydrological and biological systems, ore deposits, hydrothermal systems, etc., and some applications of non-traditional stable isotopes.

Chapter 7 is an overview of the research outcomes of the last decades and how geochemistry affects the sustainability of the environment and causes pollution at different levels. This chapter also explains the geochemical reactions of environmental pollutants and the physicochemical characteristics of various types of environmental samples.

Chapter 8 focuses on the development of geochemistry and its distinct branches, as isotopic geochemistry reveals information about environmental health to avoid harmful chemicals that have antagonistic effects on health. Also discussed is the role of medical geochemistry in addressing many environmental health problems.

Chapter 9 discusses the importance of inorganic geochemistry, the rock-forming minerals and their compositions. The characterization of minerals, elemental analysis and the determination of composition are elaborated. Clay minerals and their applications are also explained.

Chapter 10 discusses the periodic properties such as ionization energy, electronegativity, electron affinity and influence on the chemical bond formation among the elements. The distribution of elements in different layers present on the earth which depends on factors like temperature, pressure and altitude are discussed. Several subdisciplines and the scope of geochemistry are also presented.

The Editors
February 2021

1

Toxic Geogenic Contaminants in Serpentinitic Geological Systems: Occurrence, Behavior, Exposure Pathways, and Human Health Risks

Willis Gwenzi

Biosystems and Environmental Engineering Research Group, Department of Soil Science and Agricultural Engineering, University of Zimbabwe, Mt. Pleasant, Harare, Zimbabwe

Abstract

Serpentinitic ultramafic geological systems contain toxic geogenic contaminants posing significant human health risks. The current chapter presents an overview of the nature, occurrence, behavior, and human health risks of toxic geogenic contaminants in serpentinitic geological systems. The geogenic contaminants include toxic metals (Co, Cr, Fe, Mn, Ni, and Zn), chrysotile asbestos, and rare earth elements. Occupational exposure occurs via inhalation in mining, milling, sculpturing, engraving, and carving industries. Non-occupational exposure include inhalation, and ingestion of toxic geogenic contaminants in contaminated geophagic earths, wild foods, herbal medicines, and water. Human exposure risk factors in Africa include weak and poorly enforced occupational and environmental regulations, consumption of contaminated foods and water, and lack of health surveillance systems. The prevalence of geophagy, high dietary intake of iron coupled with genetic disposition also increases iron overload and its health risks among native Africans. The human health risks of chrysotile (e.g., asbestosis, cancers) and toxic metals and rare earth elements (e.g., oxidative stress) are summarized. Human health risks may also occur via synergistic interactions among toxic geogenic contaminants (e.g., chrysotile and toxic metals), and with other

Email: wgwenzi@yahoo.co.uk; wgwenzi@agric.uz.ac.zw; (ORCID: http://orcid.org.0000-0003-3149-1052)

Inamuddin, Mohd Imran Ahamed, Rajender Boddula and Tariq Altalhi (eds.) Geochemistry: Concepts and Applications, (1–22) © 2021 Scrivener Publishing LLC

health stressors (e.g., infectious diseases). Finally, mitigation measures to safeguard human health and future needs are highlighted.

Keywords: Serpentines, ultramafics, medical geology, toxic metals, chrysotile asbestos, rare earth elements, asbestosis, human exposure pathways

1.1 Introduction

Serpentinitic geological environments, consisting of serpentines and their ultramafic precursors or protoliths occur on all continents and cover about 1% of the earth surface [1]. To date, research on serpentinitic geological environments has been limited to their geology, geochemistry, and ecology [1, 2]. However, the medical geology of serpentinitic geological systems has received limited research attention. Medical geology seeks to understand how the interactions between humans and animals, and geological processes and activities influence human and animal health [3, 4]. The field of medical geology has attracted significant public and research attention, and an increasing body of literature exists on the subject [3, 5, 6]. The literature on medical geology, include (1) reviews presenting an overview of the discipline [3, 5, 6] and (2) empirical studies investigating the human health risks of specific geological processes and contaminants [7, 8].

Serpentinitic geological systems contain anomalously high concentrations of toxic geogenic contaminants, which pose human health risks. These include, toxic elements such as rare earth elements and metals, and chrysotile asbestos [3, 4, 9]. Toxic metals occurring in serpentinitic geological environments include cobalt (Co), chromium (Cr), iron (Fe), manganese (Mn), nickel (Ni), and zinc (Zn). The 17 rare earth elements include the lanthanides, Yttrium, and Scandium [10]. Unlike toxic metals and rare earth elements, chrysotile asbestos is not a chemical element, but a compound. In view of this, the term "toxic geogenic contaminants" is broadly used in this chapter to refer to both chrysotile and toxic elements. Limited reviews exist on the occurrence, human exposure, and health risks of toxic geogenic contaminants in serpentinitic geological systems. An exception is a recent review investigating the nature, biogeochemical behavior, human exposure pathways, and health risks of toxic geogenic contaminants in serpentinitic ultramafic geological systems [4].

The current chapter seeks to present an overview of the nature, occurrence, environmental behavior, and human exposure pathways and health risks associated with toxic geogenic contaminants in serpentinitic geological systems. The specific objectives are to (1) discuss the occurrence and environmental behavior of toxic geogenic contaminants, (2) summarize the human exposure and health risks of toxic geogenic, and (3) highlight mitigation measures and future research needs. Figure 1.1 summarizes the key focus on the chapter, including the nature, human exposure pathways, and health risks of toxic geogenic contaminants in serpentinitic geological systems and their ultramafic protoliths.

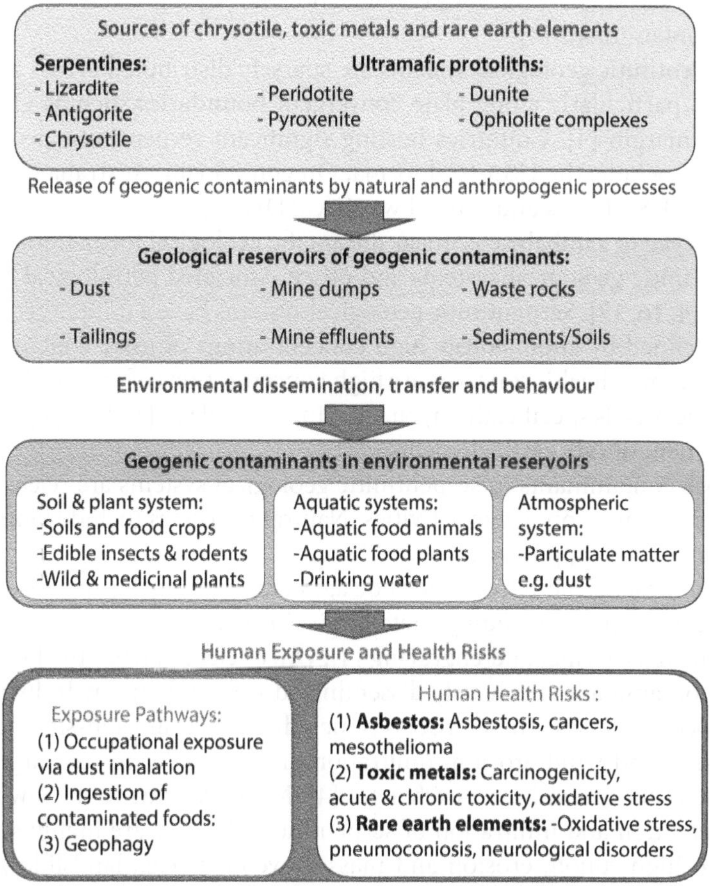

Figure 1.1 Summary of the nature, environmental sources, and human exposure and health risks of toxic geogenic contaminants in serpentinitic geological systems and their ultramafic protoliths. Adapted and redrawn based on Gwenzi [4].

1.2 Serpentinitic Geological Systems

1.2.1 Nature, Occurrence, and Geochemistry

The serpentine group consists of three minerals: (1) chrysotile, which is a type of asbestos; (2) antigorite; and (3) lizardite (www.minerals.net; www.geology.com). The hydrothermal transformation of ultramafic rocks or protoliths such pyroxenites and peridotites results in the formation of serpentines [9, 11]. In this chapter, the term serpentinitic geological systems is broadly used to refer to both serpentines *sensu stricto*, and their ultramafic protoliths or precursor rock assemblages. This is because both serpentines and their ultramafic protoliths contain anomalously high concentrations of toxic contaminants [4].

Serpentinitic geological systems are sparsely distributed across all continents, particularly along plate convergent boundaries such as Circum-Pacific margin [4]. Countries hosting significant serpentinitic geological systems include, the USA (California, Oregon), Spain, Australia (Western Australia), Sri Lanka and Zimbabwe (Great Dyke) [4, 12–15]. In Africa, the Great Dyke in Zimbabwe is linear ultramafic geological formation hosting serpentinitic geological systems and other associated petrological assemblages [4, 16, 17]. Serpentinitic geological systems have a unique geochemistry, defined by anomalously high concentrations of toxic contaminants [1]. Of human health concerns are high concentrations of chrysotile asbestos, toxic metals specifically Cr, Ni, Fe, Mn, Co, and Cd [1, 9], and potential enrichment of rare earth elements.

Toxic contaminants in serpentinitic geological systems are released via natural and anthropogenic activities. Natural processes include biotic and abiotic weathering processes, and subsequent erosion and mass movement [18]. Anthropogenic activities include agriculture (i.e., tillage practices) and mining operations including metallurgical processing, construction, and sculpturing, carving, and engraving [4]. Once released from the host rock toxic contaminants are widely disseminated via wind and hydrologically-mediated processes. These processes include surface and sub-surface run-off, water and wind erosion, infiltration and groundwater recharge, and surface water-groundwater exchanges [4]. Moreover, due to the low moisture retention and infiltration of serpentine soils, they are highly susceptible to high runoff, erosion and mass movement (e.g., landslides) [19]. Therefore, the toxic contaminants from serpentinitic geological systems tend to be widely disseminated into the surrounding environment, resulting in off-site environmental pollution and human health risks [4, 19].

1.2.2 Occurrence and Behavior of Toxic Contaminants

1.2.2.1 Chrysotile Asbestos

Chrysotile is a fibrous asbestos mineral, which is a human carcinogen [4, 20]. Due to its toxicity, chrysotile and other types of asbestos have been long-banned in most developing countries. The physico-chemical behavior of chrysotile is defined by its crystal habit, fiber size, chemical composition, biopersistence, and surface reactivity, including capacity to generate reactive oxygen radicals [4, 21, 22]. Its fibrous nature, low density, and long residence time, coupled with unique aerodynamic properties make it a unique contaminant. For example, due to its biopersistence and aerodynamic properties, chrysotile can be transported over long distances from the sources to other environmental compartments [4]. Chrysotile undergoes limited biochemical degradation and is not taken up by plants due to its fibrous nature. Evidence shows that acid rain may promote the corrosion of chrysotile, and its release into the environment. In serpentinitic geological systems, chrysotile co-occurs with other contaminants such as toxic metals [21–23], thus may act as a carrier in this regard. The co-occurrence of chrysotile and other toxic contaminants points to potential synergistic interactions between the two, a process reported to have adverse human health effects [21, 22].

A few studies exist on the environmental and biogeochemical behavior of chrysotile, particularly in aqueous systems [24, 25]. The following summary findings are evident from existing data: (1) coagulation and filtration significantly remove chrysotile in aqueous systems, including drinking water supplies [26], (2) strong organic acids promote the loss of crystallinity, and release of Mg, which alters the surface charge from an initial positive to a negative one [24, 25]. Therefore, some studies suggest that the release of organic acids from organic amendments (sawdust, peat, compost, manure, and biochar) can be used as basis to reclaim serpentinitic geological systems such as mine tailings and waste dumps [4, 24].

1.2.2.2 Toxic Metals

Serpentinitic geological systems contain extremely high concentrations of toxic metals, specifically Fe, Ni, Cr, Co, Mn, and Zn [4, 27, 28]. The total concentrations of toxic metals have been reported to be in the order of hundreds and thousands mg per kg of dry soil. For example, the following total concentration of toxic metals was reported in serpentinitic geological systems in Albania: 3,865 mg Cr/kg, 3,579 mg Ni/kg, 2,495 mg

Zn/kg, 1,107 mg Cu/kg, and 476 mg Co/kg [26]. High concentrations of toxic metals were also reported elsewhere in India and Zimbabwe [27]. Two points are noteworthy literature on toxic metals: (1) high total concentrations of Cr and Ni are considered universal in serpentines, but they co-occur with other toxic metals (e.g., Zn, Fe, and Mn) in equally high concentrations, and (2) the bulk of the data on toxic metals report total concentrations, which indicate the total pool in the soil, but provide no indication on the bioavailable and bioaccessible fractions [4]. The lack of data on bioavailable and bioaccessible concentrations makes it difficult to infer human health risks from such data. In some instances (e.g., Cr), some metal species (e.g., Cr (VI)) are more toxic than others (e.g., Cr(III)), but data on speciation are rarely provided in most studies on toxic metals.

The literature on metal geochemistry in serpentinitic geological environments is dominated by Cr, particularly Cr(VI), probably due to its high toxicity [4, 29]. Specifically, several studies exist on the role of Mn minerals (e.g., MnO_2, birnessite,) in accelerating the dissolution of Cr-bearing spinels such as chromite ($FeCr_2O_4$) and magnesiochromite ($MgCr_2O_4$), and the oxidation of Cr(III) to Cr(VI) [1, 30]. The accelerated dissolution and oxidation account for extremely high concentrations of Cr(VI) detected in aquatic ecosystem in serpentinitic geological systems [9, 29]. For instance, a recent review showed that high Cr(VI) concentrations above the World Health Organization (WHO) limit for drinking water (i.e., 50 $\mu g\ L^{-1}$ equivalent to 960 nM Cr(VI)) have been detected in both surface water and groundwater in several countries [4]. Such high concentrations may pose significant human health risks in cases where communities rely on untreated drinking water.

Unlike chrysotile, toxic metals may undergo uptake and bioaccumulation by wild plants and food crops. Rhizospheric processes including, root-soil interactions and the release of root exudates are critical in the biogeochemical behavior of toxic metals [4, 31]. For instance, root-soil interactions and root exudates may increase the bioavailability and bioaccesibility of toxic metals, and their subsequent uptake and bioaccumulation by plants and crops [31]. Such processes may transfer toxic metals from the soil system into the human food chain, thereby posing human health risks.

1.2.2.3 Rare Earth Elements

Rare earth elements are often excluded in literature on the geochemistry and medical geology of serpentinitic geological systems. Exceptions are cases where they are used as geochemical or environmental tracers [13]. Yet, evidence shows that rare earth elements may undergo enrichment in serpentinites such as peridotite and dunite, as well as their ultramafic protoliths [32, 33]. The enrichment of rare earth elements have been attributed to various processes occurring during the hydrothermal transformation [32, 34]. First, a decline in the solubility of rare earth elements is attributed to a rise in the pH of hydrothermal solutions during the transition from acidic to alkaline conditions after interacting with alkaline ultramafic rocks. Second, the interactions between the ultramafic rocks and hydrothermal fluids/melts and sea water rich in rare earth elements promote enrichment of rare earth elements. Therefore, rare earth elements may occur in high concentrations in serpentinitic geological systems, where they may undergo simultaneous release and dissemination with other toxic contaminants. The dissemination of rare earth elements may occur via contaminated sediments, and solid wastes and wastewaters from mining and mineral processing [4, 10, 35].

1.3 Human Exposure Pathways

1.3.1 Occupational Exposure

Toxic contaminants in serpentinitic geological systems may occur in various environmental compartments, including soils, wild plants, crops, and animals. Once in these environmental compartments, toxic contaminants may enter the human body via occupational and non-occupational exposure [4]. Occupational exposure to toxic contaminants may occur via inhalation in industrial production systems [36]. Typical industries promoting occupation exposure are: (1) mining and mineral processing, including quarries; (2) production of frictional materials such as brake pads, textiles, gas masks, cement, and asbestos; (3) agriculture; (4) construction; and (5) sculpturing, engraving, and carving [4, 37, 38]. Occupational exposure to chrysotile asbestos has been linked to human health risks such as kidney and ovarian cancers, respiratory diseases, and mesothelioma [38].

1.3.2 Non-Occupational Exposure Routes

1.3.2.1 Inhalation of Contaminated Particulates

Non-occupational exposure may occur via inhalation of contaminated air-borne particulates among populations living close to mining, construction, and agricultural activities [4]. For example, several abandoned chrysotile mines and waste dumps in Australia [39], South Africa [40], and

Zimbabwe [41, 42], posing non-occupational exposure risks to communities. Non-occupational exposure may also occur via the transfer of toxic contaminants from clothes of workers to their family members.

1.3.2.2 Ingestion of Contaminated Geophagic Earths

The deliberate ingestion of geophagic earths such as clays and termite mounds, which is referred to as geophagy, may contribute to non-occupational exposure. Geophagy, which is practiced for various cultural and perceived health reasons, is common in several communities in Africa [10, 43]. For example, in Kenya, it is estimated that women consume 40 g per day of geophagic earths, contributing to iron intake of at least about nine times the maximum permissible daily intake [43]. High intake of toxic metals via geophagy has also been reported in other studies [44]. Although data pertaining to geophagy in serpentinitic geological systems are still missing, the intake of toxic contaminants could be higher in such environments compared to non-serpentinitic environments. This risk could be particularly higher among pregnant women and their unborn babies. This is because pregnant women have a high intake of geophagic earths, which is perceived to reduce anemia and nausea [45].

1.3.2.3 Ingestion of Contaminated Drinking Water

Toxic contaminants including chrysotile and toxic metals such as Cr and its highly toxic form Cr(VI) have been reported in aquatic systems in serpentinitic geological environments as early as the 1980s [46–48]. Cr(VI) exceeding the maximum permissible drinking water limit of 50 μg/L WHO [49] has also been detected in several aquatic systems, including groundwater, and surface water systems [50–52]. Thus, the consumption of untreated contaminated drinking water, a common practice in most developing countries may constitute a human exposure route to toxic contaminants.

1.3.2.4 Ingestion of Contaminated Medicinal Plants

Several herbal and medicinal plants have been reported to contain toxic metals exceeding permissible limits [4, 53]. For instance, a common medicinal plant (St. John's Wort, *Hypericum perforatum* L.) growing on serpentinitic substrate had high concentrations of Cd, Ni, and Cr in dry plant material above the WHO permissible limits. Moreover, several medicinal and herbal plants in Africa (e.g., *Senecio coronatus* (Thunb.) and *Datura metal* L. (Solanaceae)) are known to be metallophytes and metal hyperaccumulators even under natural conditions [4, 54]. Hence, intake of herbal and medicinal plants constitutes a potential non-occupational exposure route especially for low income populations with limited access to modern health care.

1.3.2.5 Ingestion of Contaminated Wild Foods

Edible wild plants and animals foods such as mushrooms and honey harvested from serpentinitic geological environments have been reported to have high concentrations of toxic contaminants [4]. For example, wild edible mushroom species (e.g., *Russula delica*) harvested from serpentine had higher Cd, Cr, and Ni than those from volcanic sites [55]. Higher Al, Zn, and Pb were also observed in edible mushrooms in the Great Dyke (Zimbabwe), a well-known serpentinitic geological system than those from non-serpentinitic environment [56]. Wild honey harvested from the wild and apiaries in serpentinitic environments had high concentrations of toxic metals compared to that from the control [57, 58]. In Kosovo, the concentration of nickel in honey from serpentinitic flora (3.71 mg/kg) was twice that of the non-serpentine one (1.66 mg/kg) [58]. The same authors concluded that the high Ni in honey originated from Ni in dust from serpentine soils, and nectar collected by honeybees from Ni accumulating plants growing on serpentine soils.

As Gwenzi [4] pointed out, food crops, livestock products such as meat and milk, and edible rodents and insects derived from serpentinitic geological environments may also contain high concentrations of toxic contaminants. For example, paddy rice from serpentine soils had high total Ni concentration of 472 mg/kg and posed human health risks [59]. In Galicia (Spain), forage growing on serpentines accumulated Cr, Cu, and Ni, resulting in toxic concentrations of Ni in kidneys (1.296–1.765 mg/kg) and liver (257 mg/kg) [60]. In the same study, the concentrations of Ni and Cu in animal tissues were significantly correlated to concentrations in the soils and forage (r^2 = 0.71–0.87). Insects and rodents occurring on metal contaminated environments have been reported to accumulate toxic contaminants such as metals [61–63]. Although data on toxic contaminants in edible insects and rodents on serpentinitic geological environments are still lacking, one may infer that such edible insects and rodents may also accumulate toxic contaminants [4]. Hence, the consumption of wild foods is a non-occupational exposure route for toxic contaminants.

1.4 Human Health Risks and Their Mitigation

1.4.1 Health Risks

1.4.1.1 Chrysotile Asbestos

The human health risks of chrysotile asbestos are the most documented among the three groups of toxic contaminants occurring in serpentinitic geological systems. Chrysotile is considered as a carcinogen and has been linked to incidences of human health conditions. High incidences of asbestosis, lung and ovarian cancers, and mesothelioma have been associated with chrysotile asbestos [4, 38]. Specifically, increased incidences of cancer in human populations inhabiting serpentinitic geological environments have been reported in Calabria in Southern Italy) [21, 22]. The toxicity mechanisms and carcinogenicity of chrysotile are quite complex and depends on the physico-chemical properties of the chrysotile [64, 65]. The toxicity mechanisms include (1) breakage of the deoxyribonucleic acid or gene structure and (2) the generation of highly reactive oxygen radicals that cause severe oxidative stress [4, 66].

Controversy and misconceptions exist with respect to the toxicity of chrysotile asbestos. Some studies suggest that chrysotile is less biopersistent, thus less toxic and safer than other types of asbestos such as amphiboles [67–69]. These sentiments are largely driven by sectoral interests, and are meant to promote the production and use of asbestos [4]. For example, one of the studies claiming that chrysotile is non-toxic was funded by the Asbestos Institute, Montréal, Canada, and the Government of Québec, raising potential conflict of interest [70]. Moreover, a series of studies by Bernstein and co-workers were conducted for short exposure periods (i.e., 1 year) [67, 70, 71]. Such periods are far lower than the latent period of 40 years required between time of exposure and expression of human health outcomes [36]. In Zimbabwe, where large chrysotile deposits exist in Shabani-Mashava Mine in Zvishavane [72], the Minerals Marketing Corporation of Zimbabwe claim that "If inhaled, chrysotile or white asbestos is moved from the lung while the amphiboles persist. Therefore it is the blue asbestos, amphiboles that causes asbestosis, lung cancer etc., not chrysotile asbestos." (http://www.mmcz.co.zw/products/industrial-minerals/). This claim is in total disregard of the evidence from Zimbabwe showing profound human health risks associated with exposure to chrysotile asbestos [4, 41, 42]. For example, Cullen and Baloyi [41] showed that several human health risks, including malignant mesothelioma, morbid asbestosis, non-malignant pleural disease, and lung cancer reported in other countries were also prevalent among workers in the chrysotile asbestos industry. The fact that chrysotile is carcinogenic just like amphibole asbestos is also shared by authoritative global health agencies, including the WHO and the International Agency for Research on Cancer [4, 73, 74]. The position of the WHO and International Agency for Research on Cancer is based on comprehensive reviews of recent literature comprising of about 100 studies.

1.4.1.2 Toxic Metals

Data tracing human health outcomes to toxic metals and rare earth elements in serpentinitic geological systems remain limited. Lacking that, the human health risks of toxic metals and rare earth elements are drawn from general literature on these elements [74]. For example, toxic metals including Cr, Co, Cu, Mn, Fe, and Zn are redox active; hence, they undergo redox reactions to generate highly reactive free radical species [4, 74]. Highly reactive free radicals cause oxidative stress, which damages lipid membranes and deoxyribonucleic acids [75, 76]. Other metals such as cobalt and its salts have been associated with genotoxicity and carcinogenicity [74]. Iron is often regarded as less toxic than other heavy metals partly because it tends to occur in high background concentrations especially in tropical environments [4]. However, high intake of iron may cause iron overload and human toxicity as reported in some native African communities with a high genetic predisposition for such a conditions [77, 78]. Thus, high iron intake in iron-rich foods, drinking water, and geophagic earths derived from serpentines may increase the risk of iron overload and toxicity. Moreover, the co-occurrence of chrysotile asbestos and toxic metals may result in synergistic interactions, which may result in adverse human health outcomes [4]. For example, evidence drawn from southern Italy suggest that the high incidences of lung cancer among human populations living in serpentinitic geological environments were related to the synergistic interactions between co-occurring asbestos and the toxic metals particularly Ni and Cr [21, 22].

1.4.1.3 Rare Earth Elements

Rare earth elements are highly reactive elements with unique physicochemical properties [10]. In cases where the enrichment of rare earth elements occurs in serpentinitic geological environments, they are likely to be simultaneously taken up by humans together with other toxic contaminants. Evidence on the human health risks of rare earth elements derived from serpentinitic environments are scarce. However, the human health risks of rare earth elements are well-known, and have been reviewed in an earlier paper [10]. Examples of health risks include, damage to the human nephrological system, pneumoconiosis, dysfunctional neurological disorders, and anti-testicular effects and male sterility, among others [4, 10, 79, 80]. A detailed review of the health effects of rare earth elements were presented in earlier literature [10]. The bulk of the evidence on human health risks relate to the application of rare earth elements in agriculture, health care, and occupational exposure [10, 81]. By compassion, data relating to serpentinitic geological systems are scarce, highlighting the need for further research on this aspect.

1.4.2 Mitigating Human Exposure and Health Risks

Environmental and human health risk assessment and mitigation are required to safeguard human health against toxic contaminants in serpentinitic geological systems. A detailed discussion of the human health risk assessment protocol and mitigation for toxic contaminants in serpentinitic geological environments is presented in an earlier review [4]. Here, an overview of the risk analysis, evaluation, and mitigation, including specific interventions, is presented based on literature [4, 10].

1.4.2.1 Risk Analysis

This step entails the identification and characterization of the nature of human exposure and health risks. Key activities include (1) determining the concentrations of the toxic contaminants, (2) identifying the human population at risk, (3) determining exposure pathways, and (4) estimating daily intakes of toxic contaminants relative to permissible maximum guidelines.

1.4.2.2 Risk Evaluation

Risk evaluation involves the determination of the likelihood of occurrence and the associated human health consequences of exposure to toxic contaminants. A risk evaluation framework based on a combination of likelihood and consequences can be used to rank the risks as follows: "extremely high", "high", "moderate", and "low/negligible". Such a framework can be used to identify human health risks warranting mitigation and for prioritization and allocation of scarce resources.

1.4.2.3 Risk Mitigation

Risk mitigation involves identifying the mitigation interventions and evaluating their potential to address the human health risks, and in terms of feasibility and cost. This is then followed by the implementation of the mitigation strategy, and subsequent monitoring and evaluation and feedback to improve performance.

1.4.2.4 Overview of Mitigation Interventions

Several mitigation measures have been highlighted for safeguarding human health in serpentinitic geological environments [4]. These include (1) proper land use planning to avoid human settlement in hotspot areas, (2) the use of appropriate barriers and safety procedures (e.g., special respirators) to isolate humans from occupational exposure to toxic contaminants, (3) soil conservation practices and use of soil amendments such as biochar to reduce erosion, mobility, and dissemination of toxic contaminants [82, 83]; (4) restoration and stabilization of post-mining landscapes to reduce emissions of toxic contaminants [84], (5) the application of low-cost methods for the removal of toxic contaminants in drinking water sources such as biochar-based filters [85], and (6) in the case of chrysotile asbestos, the development and use of alternative substitutes which are non-toxic in order to ultimately stop the mining, processing, and application of chrysotile [4, 36]. In most cases, depending on the severity of the human exposure and health risks, a combination of several interventions may be required to safeguard human health.

1.5 Future Perspectives

The medical geology of serpentines is an emerging topic of research that requires an integration of earth sciences, public health and environmental sciences. As pointed out in an earlier paper, further research is required to address several knowledge gaps [4]. Addressing these knowledge gaps is key to risk assessment and mitigation of the human health risks associated with serpentine geological systems.

(1) Understanding the occurrence and spatial distribution of serpentines in various countries hosting serpentinitic geological environments through the application of geostatistical spatial analysis and remote sensing.

(2) Detailed understanding of the behavior and fate processes of toxic contaminants, including their speciation and partitioning into various phases through mass balance analysis and speciation modelling.

(3) Understanding the human toxicology and ecotoxicology of toxic contaminants, including determining key exposure routes, bioavailability and bioaccesibility, and daily intakes using established protocols for human health risk assessment.

(4) Comprehensive understanding of the relationship between toxic contaminants and human health outcomes through case-control epidemiological studies, particularly in Africa.

(5) Establishing environmental and human health surveillance systems to determine baseline conditions and current status of human and environmental health in serpentinitic geological environments.

1.6 Conclusions

Understanding the nature and human exposure and health risks of toxic geogenic contaminants in serpentinitic ultramafic geological systems is critical for safeguarding human health. The current chapter presented an overview of the nature, occurrence, environmental behavior, and human exposure and health risks associated with toxic geogenic contaminants in serpentinitic ultramafic geological environments. Toxic geogenic contaminants of human health concern include toxic metals, rare earth elements, and chrysotile asbestos. Occupational and non-occupational exposure are the dominant human intake pathways. Human occupational exposure via inhalation occurs in industrial settings such as mining, milling, sculpturing, engraving, and carving. Non-occupational exposure pathways include inhalation of toxic geogenic contaminants and ingestion of contaminated geophagic earths, wild foods, herbal medicines, and water. Weak and poorly enforced occupation and environmental regulations, lack of environmental and human health surveillance systems coupled with high consumption of potentially contaminated wild foods, geophagic earths, and water increase human exposure and health risks in Africa. The prevalence of geophagy and high dietary intake of iron coupled with genetic disposition increases iron overload and the associated human health risks among native Africans. The human health risks of chrysotile include asbestosis, cancers, and mesothelioma, while toxic metal and rare earth elements induce oxidative stress which damages biomolecules such as deoxyribonucleic acid. Human health risks may also occur via synergistic interactions among toxic geogenic contaminants (e.g., chrysotile and toxic metals), and even between toxic geogenic contaminants and other health stressors such as the prevalence of infectious diseases. Mitigation measures to safeguard human health, and future research directions were highlighted.

Acknowledgements

This chapter is based on research project entitled, "*The Potential of Native Plants of the Ultramafic Great Dyke of Zimbabwe for the Phytoremediation and Restoration of Metalliferrous Mine Wastes*", which was funded by the British Ecological Society (BES) Ecologists in Africa Grant No. 5774-6818. I am very grateful for the financial support provided by the British Ecological Society (BES) Ecologists in Africa. However, I am solely responsible for the views expressed in this chapter, and the decision to publish the research, BES, and its affiliates played no role whatsoever in the research and decision to publish the chapter.

References

1. Oze, C., Skinner, C., Schroth, A.W., Coleman, R.G., Growing up green on serpentine soils: biogeochemistry of serpentine vegetation in the Central Coast Range of California. *Appl. Geochem.*, 23, 12, 3391–3403, 2008.
2. Alexander, E.B., Coleman, R.G., Harrison, S.P., Keeler-Wolfe, T., *Serpentine geoecology of western North America: geology, soils, and vegetation*, Oxford University Press, Oxford, 2007.
3. Bundschuh, J., Maity, J.P., Mushtaq, S., Vithanage, M., Seneweera, S., Schneider, J., Bhattacharya, P., Khan, N.I., Hamawand, I., Guilherme, L.R., Reardon-Smith, K., Medical geology in the framework of the sustainable development goals. *Sci. Total Environ.*, 581, 87–104, 2017.
4. Gwenzi, W., Occurrence, behaviour, and human exposure pathways and health risks of toxic geogenic contaminants in serpentinitic ultramafic geological environments (SUGEs): A medical geology perspective. *Sci. Total Environ.*, 700, 134622, https://doi.org/10.1016/j.scitotenv.2019.134622, 2020.
5. Davies, B.E., Bowman, C., Davies, T.C., Selinus, O., Medical geology: Perspectives and prospects, in: *Essentials of medical geology*, pp. 1–13, Springer, Dordrecht, 2013.
6. Goovaerts, P., Geostatistics: a common link between medical geography, mathematical geology, and medical geology. *J. South. Afr. Inst. Min. Metall.*, 114, 8, 605–613, 2014.

7. Doocy, S., Daniels, A., Packer, C., Dick, A., Kirsch, T.D., The human impact of earthquakes: a historical review of events 1980-2009 and systematic literature review. *PLoS Curr.*, 5, 2013, https://currents.plos.org/disasters/index.html%3Fp=6639.html.
8. Kut, K.M.K., Sarswat, A., Srivastava, A., Pittman Jr., C.U., Mohan, D., A review of fluoride in African groundwater and local remediation methods. *Groundwater Sustainable Dev.*, 2, 190–212, 2016.
9. Oze, C., Fendorf, S., Bird, K.D., Coleman, G.R., Chromium geochemistry in serpentinized ultramafic rocks and serpentine soils from the Franciscan complex of California. *Am. J. Sci.*, 304, 67–101, 2004.
10. Gwenzi, W., Mangori, L., Danha, C., Chaukura, N., Dunjana, N., Sanganyado, E., Sources, behaviour, and environmental and human health risks of high-technology rare earth elements as emerging contaminants. *Sci. Total Environ.*, 636, 299–313, 2018.
11. Sleep, N.H., Meibom, A., Fridriksson, T., Coleman, R.G., Bird, D.K., H2-rich fluids from serpentinization: geochemical and biotic implications. *Proc. Natl. Acad. Sci.*, 101, 35, 12818–12823, 2004.
12. Rajapaksha, A.U., Vithanage, M., Oze, C., Bandara, W.M.A.T., Weerasooriya, R., Nickel and manganese release in serpentine soil from the Ussangoda Ultramafic Complex, Sri Lanka. *Geoderma*, 189, 1–9, 2012.
13. Manning, A.H., Mills, C.T., Morrison, J.M., Ball, L.B., Insights into controls on hexavalent chromium in groundwater provided by environmental tracers, Sacramento Valley, California, USA. *Appl. Geochem.*, 62, 186–199, 2015.
14. Blades, M.L., Foden, J., Collins, A.S., Alemu, T., Woldetinsae, G., The origin of the ultramafic rocks of the Tulu Dimtu Belt, western Ethiopia–do they represent remnants of the Mozambique Ocean? *Geol. Mag.*, 156, 62–82, 1–21, 2017.
15. Antoniadis, V., Golia, E.E., Liu, Y.T., Wang, S.L., Shaheen, S.M., Rinklebe, J., Soil and maize contamination by trace elements and associated health risk assessment in the industrial area of Volos, Greece. *Environ. Int.*, 124, 79–88, 2019.
16. Oberthür, T., Davis, D.W., Blenkinsop, T.G., Höhndorf, A., Precise U-Pb mineral ages, Rb-Sr and Sm-Nd systematics for the Great Dyke, Zimbabwe - constraints on crustal evolution and metallogenesis of the Zimbabwe Craton. *Precambrian Res.*, 113, 293–306, 2002.
17. Stribrny, B., Wellmer, F.-W., Burgath, K.-P., Oberthür, T., Tarkian, M., Pfeiffer, T., Unconventional PGE occurrences and PGE mineralization in the Great Dyke: metallogenic and economic aspects. *Miner. Deposita*, 35, 260–281, 2000.
18. Morrison, J.M., Goldhaber, M.B., Mills, C.T., Breit, G.N., Hooper, R.L., Holloway, J.A.M., Diehl, S.F., Ranville, J.F., Weathering and transport of chromium and nickel from serpentinite in the coast range ophiolite to the Sacramento Valley, California, USA. *Appl. Geochem.*, 61, 72–86, 2015.

19. Alexander, E.B. and DuShey, J., Topographic and soil differences from peridotite to serpentinite. *Geomorphology*, 135, 271–276, 2011.
20. Oury, T.D., Sporn, T.A., Roggli, V.L. (Eds.), *Pathology of asbestos-associated diseases*, p. 357, Springer, Berlin, 2014.
21. Bloise, A., Barca, D., Gualtieri, A.F., Pollastri, S., Belluso, E., Trace elements in hazardous mineral fibres. *Environ. Pollut.*, 216, 314–323, 2016a.
22. Bloise, A., Punturo, R., Catalano, M., Miriello, D., Cirrincione, R., Naturally occurring asbestos (NOA) in rock and soil and relation with human activities: the monitoring example of selected sites in Calabria (southern Italy). *Ital. J. Geosci.*, 135, 2, 268–279, 2016b.
23. Gualtieri, A.F., Lusvardi, G., Zoboli, A., Di Giuseppe, D., Gualtieri, M.L., Biodurability and release of metals during the dissolution of chrysotile, crocidolite and fibrous erionite. *Environ. Res.*, 171, 550–557, 2019.
24. Holmes, E.P., Wilson, J., Schreier, H., Lavkulich, L.M., Processes affecting surface and chemical properties of chrysotile: Implications for reclamation of asbestos in the natural environment. *Can. J. Soil Sci.*, 92, 1, 229–242, 2012.
25. Holmes, E.P. and Lavkulich, L.M., The effects of naturally occurring acids on the surface properties of chrysotile asbestos. *J. Environ. Sci. Health, Part A*, 49, 12, 1445–1452, 2014.
26. Bales, R.C., *Surface chemical and physical behavior of chrysotile asbestos in natural waters and water treatment*, Keck Laboratories of Environmental Engineering Science, Division of Engineering and Applied Science, California Institute of Technology, Pasadena, CA. Report No. AC-8-84, 1984.
27. Bangira, C., Deng, Y., Loeppert, R.H., Hallmark, C.T., Stucki, J.W., Soil mineral composition in contrasting climatic regions of the Great Dyke, Zimbabwe. *Soil Sci. Soc. Am. J.*, 75, 6, 2367–2378, 2011.
28. Shallari, S., Schwartz, C., Hasko, A., Morel, J.L., Heavy metals in soils and plants of serpentine and industrial sites of Albania. *Sci. Total Environ.*, 209, 2–3, 133–142, 1998.
29. Vengosh, A., Coyte, R., Karr, J., Harkness, J.S., Kondash, A.J., Ruhl, L.S., Merola, R.B., Dywer, G.S., Origin of hexavalent chromium in drinking water wells from the piedmont aquifers of North Carolina. *Environ. Sci. Technol. Lett.*, 3, 12, 409–414, 2016.
30. Namgung, S., Kwon, M.J., Qafoku, N.P., Lee, G., $Cr(OH)_3(s)$ oxidation induced by surface catalyzed Mn (II) oxidation. *Environ. Sci. Technol.*, 48, 18, 10760–10768, 2014.
31. Dotaniya, M.L. and Meena, V.D., Rhizosphere effect on nutrient availability in soil and its uptake by plants: a review. *Proc. Natl. Acad. Sci. India Sect. B: Biol. Sci.*, 85, 1, 1–12, 2015.
32. Paulick, H., Bach, W., Godard, M., De Hoog, J.C.M., Suhr, G., Harvey, J., Geochemistry of abyssal peridotites (Mid-Atlantic Ridge, 15820'N, ODP Leg

209): Implications for fluid/rock interaction in slow spreading environments. *Chem. Geol.*, 234, 179–210, 2006.

33. Kodolányi, J., Pettke, T., Spandler, C., Kamber, B.S., Gméling, K., Geochemistry of ocean floor and fore-arc serpentinites: constraints on the ultramafic input to subduction zones. *J. Petrol.*, 53, 2, 235–270, 2011.
34. Scambelluri, M., Rampone, E., Piccardo, G.B., Fluid and element cycling in subducted serpentinite: a trace-element study of the Erro–Tobbio high-pressure ultramafites (Western alps, NW Italy). *J. Petrol.*, 42, 1, 55–67, 2001.
35. Paulick, H. and Machacek, E., The global rare earth element exploration boom: an analysis of resources outside of China and discussion of development perspectives. *Res. Policy*, 52, 134–153, https://doi.org/10.1016/j.resourpol.2017.02.002, 2017.
36. WHO (World Health Organization), *Chrysotile asbestos*, World Health Organization, Geneva, 2014.
37. Davies, T.C. and Mundalamo, H.R., Environmental health impacts of dispersed mineralisation in South Africa. *J. Afr. Earth Sci.*, 58, 4, 652–666, 2010.
38. Camargo, M.C., Stayner, L.T., Straif, K., Reina, M., Al-Alem, U., Demers, P.A., Landrigan, P.J., Occupational exposure to asbestos and ovarian cancer: a meta-analysis. *Environ. Health Perspect.*, 119, 9, 1211–1217, 2011.
39. Hendrickx, M., Naturally occurring asbestos in eastern Australia: a review of geological occurrence, disturbance and mesothelioma risk. *Environ. Geol.*, 57, 4, 909–926, 2009.
40. Braun, L. and Kisting, S., Asbestos-related disease in South Africa: the social production of an invisible epidemic. *Am. J. Public Health*, 96, 8, 1386–1396, 2006.
41. Cullen, M.R. and Baloyi, R.S., Chrysotile asbestos and health in Zimbabwe: I. Analysis of miners and millers compensated for asbestos-related diseases since independence (1980). *Am. J. Ind. Med.*, 19, 2, 161–169, 1991.
42. Cullen, M.R., Lopez-Carrillo, L., Alli, B., Pace, P.E., Shalat, S.L., Baloyi, R.S., Chrysotile asbestos and health in Zimbabwe: II. Health status survey of active miners and millers. *Am. J. Ind. Med.*, 19, 2, 171–182, 1991.
43. Miller, J.D., Collins, S.M., Omotayo, M., Martin, S.L., Dickin, K.L., Young, S.L., Geophagic earths consumed by women in western Kenya contain dangerous levels of lead, arsenic, and iron. *Am. J. Hum. Biol.*, 30, 4, e23130, 2018.
44. Odongo, A.O., Moturi, W.N., Mbuthia, E.K., Heavy metals and parasitic geohelminths toxicity among geophagous pregnant women: a case study of Nakuru Municipality, Kenya. *Environ. Geochem. Health*, 38, 1, 123–131, 2016.
45. Young, S.L., Sherman, P.W., Lucks, J.B., Pelto, G.H., Why on earth?: Evaluating hypotheses about the physiological functions of human geophagy. *Q. Rev. Biol.*, 86, 2, 97–120, 2011.
46. Toft, P., Wigle, D., Meranger, J.C., Mao, Y., Asbestos and drinking water in Canada. *Sci. Total Environ.*, 18, 77–89, 1981.

47. Bales, R.C. and Morgan, J.J., Dissolution kinetics of chrysotile at pH 7 to 10. *Geochim. Cosmochim. Acta*, 49, 11, 2281–2288, 1985a.
48. Bales, R.C. and Morgan, J.J., Surface charge and adsorption properties of chrysotile asbestos in natural waters. *Environ. Sci. Technol.*, 19, 12, 1213–1219, 1985b.
49. WHO (World Health Organization), *Chromium in drinking water. Background document for preparation of WHO guidelines for drinking water quality*, World Health Organization, Geneva, 2003.
50. Chrysochoou, M., Theologou, E., Bompoti, N., Dermatas, D., Panagiotakis, I., Occurrence, origin and transformation processes of geogenic chromium in soils and sediments. *Curr. Pollut. Rep.*, 2, 4, 224–235, https://doi.org/10.1007/s40726-016-0044-2, 2016.
51. McClain, C.N. and Maher, K., Chromium fluxes and speciation in ultramafic catchments and global rivers. *Chem. Geol.*, 426, 135–57, 2016.
52. Binda, G., Pozzi, A., Livio, F., Piasini, P., Zhang, C., Anomalously high concentration of Ni as sulphide phase in sediment and in water of a mountain catchment with serpentinite bedrock. *J. Geochem. Explor.*, 190, 58–68, 2018.
53. Pavlova, D., Karadjova, I., Krasteva, I., Essential and toxic element concentrations in *Hypericum perforatum*. *Aust. J. Bot.*, 63, 2, 152–158, 2015.
54. Street, R.A., Heavy metals in medicinal plant products - An African perspective. *S. Afr. J. Bot.*, 82, 67–74, 2012.
55. Aloupi, M., Koutrotsios, G., Koulousaris, M., Kalogeropoulos, N., Trace metal contents in wild edible mushrooms growing on serpentine and volcanic soils on the island of Lesvos, Greece. *Ecotoxicol. Environ. Saf.*, 78, 184–194, 2012.
56. Nharingo, T., Ndumo, T., Moyo, M., Human health risks due to heavy metals through consumption of wild mushrooms from Macheke forest, Rail Block forest and Muganyi communal lands in Zimbabwe. *Environ. Monit. Assess.*, 187, 738, 1–11, 2015.
57. Atanassova, J., Pavlova, D., Lazarova, M., Yurukova, L., Characteristics of honey from serpentine area in the Eastern Rhodopes Mt., Bulgaria. *Biol. Trace Elem. Res.*, 173, 1, 247–258, 2016.
58. Salihaj, M. and Bani, A., The nickel content in honey derived from serpentine and non-serpentine areas of Kosovo. *Albanian J. Agric. Sci.*, 557–563, 2017, https://www.researchgate.net/profile/Aida_Bani/publication/338825224_The_nickel_content_in_honey_derived_from_serpentine_and_non-serpentine_areas_of_Kosovo/links/5e2c77c492851c3aaddabf00/The-nickel-content-in-honey-derived-from-serpentine-and-non-serpentine-areas-of-Kosovo.pdf.
59. Hseu, Z.Y. and Lai, Y.J., Nickel accumulation in paddy rice on serpentine soils containing high geogenic nickel contents in Taiwan. *Environ. Geochem. Health*, 39, 6, 1325–1334, 2017.
60. Miranda, M., Benedito, J.L., Blanco-Penedo, I., López-Lamas, C., Merino, A., López-Alonso, M., Metal accumulation in cattle raised in a

serpentine-soil area: relationship between metal concentrations in soil, forage and animal tissues. *J. Trace Elem. Med. Biol.*, 23, 3, 231–238, 2009.

61. Martiniaková, M., Omelka, R., Jančová, A., Stawarz, R., Formicki, G., Concentrations of selected heavy metals in bones and femoral bone structure of bank (*Myodes glareolus*) and common (*Microtus arvalis*) voles from different polluted biotopes in Slovakia. *Arch. Environ. Contam. Toxicol.*, 60, 3, 524–532, 2011.
62. Azam, I., Afsheen, S., Zia, A., Javed, M., Saeed, R., Sarwar, M.K., Munir, B., Evaluating insects as bioindicators of heavy metal contamination and accumulation near industrial area of Gujrat, Pakistan. *BioMed Res. Int.*, 942751, 1–11, https://doi.org/10.1155/2015/942751, 2015.
63. Kouřimská, L. and Adámková, A., Nutritional and sensory quality of edible insects. *NFS J.*, 4, 22–26, 2016.
64. Kamp, D.W., Asbestos-induced lung diseases: an update. *Transl. Res.*, 153, 4, 143–152, 2009.
65. Boulanger, G., Andujar, P., Pairon, J.C., Billon-Galland, M.A., Dion, C., Dumortier, P., Brochard, P., Sobaszek, A., Bartsch, P., Paris, C., Jaurand, M.C., Quantification of short and long asbestos fibers to assess asbestos exposure: a review of fiber size toxicity. *Environ. Health*, 13, 1, 59, http://www.ehjournal.net/content/13/1/59, 2014.
66. Røe, O.D. and Stella, G.M., Malignant Pleural Mesothelioma: History, Controversy, and Future of a Man-Made Epidemic, in: *Asbestos and Mesothelioma*, pp. 73–101, Springer, Chatham, 2017.
67. Bernstein, D.M., Rogers, R., Smith, P., The biopersistence of Canadian chrysotile asbestos following inhalation: final results through 1 year after cessation of exposure. *Inhalation Toxicol.*, 17, 1, 1–14, 2005.
68. Hodgson, J.T. and Darnton, A., Mesothelioma risk from chrysotile. Comment on "Lung cancer mortality and fibre exposures among North Carolina asbestos textile workers". *Occup. Environ. Med.*, 67, 6, 432, 2010.
69. Wang, X., Lin, S., Yano, E., Qiu, H., Igtanius, T.S., Tse, L., Lan, Y., Wang, M., Mortality in a Chinese chrysotile miner cohort. *Int. Arch. Occup. Environ. Health*, 85, 4, 405–412, 2012.
70. Bernstein, D.M. and Hoskins, J.A., The health effects of chrysotile: current perspective based upon recent data. *Regul. Toxicol. Pharm.*, 45, 3, 252–264, 2006.
71. Bernstein, D.M., The health risk of chrysotile asbestos. *Curr. Opin. Pulm. Med.*, 20, 4, 366–370, 2014.
72. ZGS (Zimbabwe Geological Survey), Minerals of Zimbabwe. Available at: http://www.mines.gov.zw/?q=minerals-zimbabwe Accessed on 3 January 2019, 2019.
73. IARC (International Agency for Research on Cancer), *A review of human carcinogens. Part C: Arsenic, metals, fibres, and dusts*, IARC Working Group on the Evaluation of Carcinogenic Risks to Humans, Lyon, 2009.

74. IARC (International Agency for Research on Cancer), *Arsenic, metals, fibres, and dusts. IARC Monographs on the Evaluation of Carcinogenic Risks to Humans*, p. 11, IARC Working Group on the Evaluation of Carcinogenic Risks to Humans, 100(PT C), Lyon, 2012.
75. Papanikolaou, G. and Pantopoulos, K., Iron metabolism and toxicity. *Toxicol. Appl. Pharmacol.*, 202, 2, 199–211, 2005.
76. Puntarulo, S., Iron, oxidative stress and human health. *Mol. Aspects Med.*, 26, 4-5, 299–312, 2005.
77. Moyo, V.M., Mandishona, E., Hasstedt, S.J., Gangaidzo, I.T., Gomo, Z.A., Khumalo, H., Saungweme, T., Kiire, C.F., Paterson, A.C., Bloom, P., MacPhail, A.P., Evidence of genetic transmission in African iron overload. *Blood*, 91, 3, 1076–1082, 1998.
78. Gordeuk, V.R., African iron overload. *Semin. Hematol.*, 39, 4, 263–269, 2002.
79. Marzec-Wróblewska, U., Kaminski, P., Lakota, P., Ludwikowski, G., Szymanski, M., Wasilow, K., Stuczynski, T., Bucinski, A., Jerzak, L., Determination of rare earth elements in human sperm and association with semen quality. *Arch. Environ. Contam. Toxicol.*, 69, 2, 191–201, 2015.
80. Pagano, G., Guida, M., Tommasi, F., Oral, R., Health effects and toxicity mechanisms of rare earth elements—Knowledge gaps and research prospects. *Ecotoxicol. Environ. Saf.*, 115, 40–48, 2015.
81. Vergauwen, E., Vanbinst, A.M., Brussaard, C., Janssens, P., De Clerck, D., Van Lint, M., Houtman, A.C., Michel, O., Keymolen, K., Lefevere, B., Bohler, S., Central nervous system gadolinium accumulation in patients undergoing periodical contrast MRI screening for hereditary tumor syndromes. *Hered. Cancer Clin. Pract.*, 16, 1, 2, 2018.
82. Bandara, T., Herath, I., Kumarathilaka, P., Hseu, Z.Y., Ok, Y.S., Vithanage, M., Efficacy of woody biomass and biochar for alleviating heavy metal bioavailability in serpentine soil. *Environ. Geochem. Health*, 39, 2, 391–401, 2017.
83. Kumarathilaka, P. and Vithanage, M., Influence of *Gliricidia sepium* biochar on attenuate perchlorate-induced heavy metal release in serpentine soil. *J. Chem.*, Article ID 6180636, 1–9, https://doi.org/10.1155/2017/6180636, 2017.
84. Hamilton, J.L., Wilson, S.A., Morgan, B., Turvey, C.C., Paterson, D.J., Jowitt, S.M., McCutcheon, J., Southam, G., Fate of transition metals during passive carbonation of ultramafic mine tailings via air capture with potential for metal resource recovery. *Int. J. Greenhouse Gas Control*, 71, 155–167, 2018.
85. Gwenzi, W., Chaukura, N., Noubactep, C., Mukome, F.N.D., Biochar-based water treatment systems as a potential low-cost and sustainable technology for clean water provision. *J. Environ. Manage.*, 197, 732–749, 2017.

2

Benefits of Geochemistry and Its Impact on Human Health

Abel Inobeme[1], Charles Oluwaseun Adetunji[2*], Muhammad Akram[3], Maliki Munirat[1], Inamuddin[6], Umme Laila[3], S.O. Okonkwo[4], Saher Islam[7] and Jonathan Inobeme[5]

[1]Department of Chemistry, Edo University Iyamho, Iyamho, Nigeria
[2]Applied Microbiology, Biotechnology and Nanotechnology Laboratory, Department of Microbiology, Edo University Iyamho, Auchi, Edo State, Nigeria
[3]Department of Eastern Medicine, Government College University Faisalabad, Faisalabad, Pakistan
[4]Department of Chemistry, Osaka Kyoiku University, Osaka, Japan
[5]Department of Geography, Ahmadu Bello University, Zaria, Nigeria
[6]Department of Chemistry, King Abdulaziz University, Jeddah, Saudi Arabia
[7]Institute of Biochemistry and Biotechnology, Faculty of Biosciences, University of Veterinary and Animal Sciences, Lahore, Pakistan

Abstract

Globally, mankind is suffering from numerous diseases and nutritional deficiencies. The rapid increase in the population of the whole world has aggravated this scenario. The population of mankind has been stipulated to increase tremendously to 9 billion in the year 2050. Some of the challenges currently experience by majority of the global population includes suffering from several diseases, nutritional deficiencies, tropical diseases, etc. Therefore, there is a need to harness all the various natural resources available in the soil and rocks, which are present in every part of the world. Proper understanding of the benefits and demerits of these essential elements might be a proper way in establishing the role in alleviating several health challenges, nutritional disorderliness, and their application in the management of human diseases. Therefore, this chapter intends to report detailed information on the application of these essential minerals for the treatment of human diseases, like dental problem, cancer, hypertension, cardiovascular

**Corresponding author*: charliguitar@yahoo.com; adetunjicharles@gmail.com; adetunji.charles@edouniversity.edu.ng; (ORCID: 0000-0003-3524-6441)

Inamuddin, Mohd Imran Ahamed, Rajender Boddula and Tariq Altalhi (eds.) Geochemistry: Concepts and Applications, (23–36) © 2021 Scrivener Publishing LLC

diseases (CVDs), goiter, and many more. Special attention was also laid on the application of these essential elements for the management of animal and plant diseases. Also, special highlights were made on some factor like temperature, rainfall, and hardness of water that affect the geochemical nature.

Keywords: Geochemistry, human health, nutritional disorderliness, dental problem, cancer, hypertension

2.1 Introduction

The pattern of dissemination of mineral elements in the crust of the Earth is not spontaneous but is regulated by physico-chemical processes, which are becoming better understood due to geochemical development [1]. Knowledge of the geochemical distribution of elements involves elucidating the comparative and unlimited abundance of the chemical elements on Earth and in its various sections—the crust, hydrophere interior, and atmosphere. It is a large part of the geochemical research, which deals with space and time distribution of the chemical elements and the rules governing it allocation [2]. Geochemistry is a unique discipline that deals with the learning that involves deposits of the mineral. Mineralization includes process that are chemical in nature that could eventually result in metal precipitation or formation of minerals. Therefore, it is essential to perform research that entails geochemical features of mineral deposits. It deals with the complex interactions between the structures of rock-soil-water-air life, which display marked variation from place to place [1]. Depending on the mineral, minerals basically refer to inorganic nutrients that are typically needed in varying amounts from lower to higher amount than 1 to 2,500 mg per day. Typical examples of this include the application of calcium to build and sustain bone and normal nervous and muscular function [3]. Multiple micronutrient powder containing at least vitamin A, iron, and zinc were added to the 2019 WHO List of Essential Medicines [4]. At the 1990 World Children's Summit, the countries gathered described deficiencies in 2 micro minerals and micronutrient (iron, iodine, and vitamin A) as being primarily widespread and presenting health risks in developing countries [5].

A mineral is an inorganic element or compound which occurs naturally and has an internal structure with unique chemical composition, crystal shape, and physical properties. Minerals such as gold are metallic or nonmetallic, for example, talc [6]. A mineral occurrence is a concentration of a mineral that is considered important by someone else or that is of scientific or technological value (usually regarded in terms of some material, like gold). In extreme cases, the product cannot even be concentrated above its normal abundance of crystals [6]. Therefore, this chapter will provide holistic information on the application of these essential mineral for the treatment of human diseases. The merits and the demerits of these essential elements were highlighted. Further suggestions and recommendation that could led to extensive research on the application of geochemistry majorly that focuses on unreported essential mineral and their practical utilization for the management of several human diseases.

2.2 General Overview of Geochemistry and Human Health

The function of geo-chemistry especially geochemistry which deals with environmental sector is very important because it deals with elements geographical distribution and gave the information about different studies including agriculture, fertility of soil, mineral exploration, nutrition, pollution and animal health, and most important human health [7]. Other than in geochemistry, we also discuss the presence of trace element and their role in different field [8]. Some elements play significant role in human health. In geochemistry, mostly, we discuss the distribution of elements on earth crust or earth surface and then researcher note the result of these elements either it is good for human health or not [9]. Some trace element effective for human health or some shows toxic effect, these elements also affect the fertility of soil [10]. Human health mostly affected when toxic or more amount of element is taking in the form of water or food and from atmosphere. Many trace elements like iodine and fluorine and many more are essential for human health, e.g., by the deficiency of iodine many health-related problems like goiter occur, due to fluoride deficiency dental problem occurs. The health-related problem in human mostly correlate with soil and due to this either man suffered with specific-illness or cause general health issue. Soil related problem mostly occur in those individuals who used soil for eating purpose; then, it causes health issue because it contains some toxic element and when people inhale these elements then it causes toxic effect on health. Nowadays, this health-related issue in human, animal, and plants which are affected through soil takes attention of web. If the level of these elements in soil increase or decrease, then it causes health related problem and affected the production of crops and agriculture, e.g., some areas have low level of trace elements like rocks of arenaceous and granite. For the investigation of human health and application of agriculture and ecology, the role of regional geo-chemical is very important. These trace elements cause many nutritional and health related problem like cancer, HTN mostly occurs through cadmium, multiple sclerosis cause due to lead and CVD problem occurs due to hardness of water. The chances of cancer mostly occur in those individuals who worked in industry then it affective through the environment, air, water, and lifestyle. The role of geo-chemistry is very important in both human health and agriculture [11].

2.2.1 Types of Geochemistry

Geochemistry divided into different types include: regional geo-chemistry, isotope geo-chemistry, organic geo-chemistry, photo geo-chemistry, cosmo-chemistry, biogeo-chemistry, and aqueous geo-chemistry.

2.2.2 Some Beneficial Effect of Some Mineral With Health Benefits

The nutritional value of the mineral elements cannot be overemphasized in animals, humans, and plants. The incorporation of mineral element inside animal feed has been shown to improve the process involved metabolic activities of the animal. Livestock grazing usually receives no mineral supplement except sodium chloride and must rely almost entirely on forage for their mineral requirements most especially from tropical regions [3]. Most metals, including copper pipes or iron skillets, are used for making solid and sturdy everyday items. But, in our bodies, they do not form such solid and lasting artifacts. Instead, it takes a lot of vital metals to activate enzymes—molecules with important body jobs. Metals do have many other important positions. Regular intakes of many minerals are needed for the human body to continue its basic functioning. The minerals elements such as zinc, calcium, magnesium, copper, and manganese are considered to be essential for the appropriate functioning and development of the well-being of the living organism. Many other criteria are for trace minerals [12]. Mineral deficiencies may be caused by low-quality feed, poor body absorption, or assimilation or result to higher request for minerals element during lactation, intense development, and pregnancy. Essential minerals—that is, those required for human health—are divided into two equally important groups: main minerals and trace minerals [13]. The main minerals used and processed in human body are calcium, chloride, magnesium, arsenic, sulphur, potassium, and sodium. Trace minerals are just as important to our well-being as the big minerals, but we do not need huge quantities. This group of minerals includes chromium, copper, fluoride, fluorine, molybdenum, manganese, iron, selenium, and zinc. Some valuable mineral elements and their health relevance are discussed as follows.

2.2.2.1 Magnesium

It has been observed that long-term chronic consumption of magnesium can manifest as a hidden magnesium deficiency showing visible signs such as cramps and weakness of muscles. A strategy was present to enhance magnesium ingestion so as to prevent heart-related complications; this has led to a gradual decline in the death rate. Simple and inexpensive to restore and maintain sufficient magnesium store [14].

2.2.2.2 Manganese

Manganese plays a vital role in many ways most especially in enzyme activation which is required for the assimilation and absorption of nutrients and foods. It is also useful in the production of a younger progeny and development of bones. This is often described as the mineral "brain", because mental function is essential. Manganese shortage is very intermittent, and difficult to detect. Nevertheless, numerous individuals cannot get the optimal rates required for safety. Low manganese is most usually caused by inadequate dietary consumption, either because of a nourishment deficient in manganese nutrition sources or because of abdominal tract illnesses that restrict nutrient intake from food. Because of the role of manganese in enzyme activation, many bio-chemical processes inside the body can be adversely affected by a deficiency but can be difficult to define. Since energy and heat are so important for the body, and the person might experience tiredness and general weakness [12].

2.2.2.3 Calcium

Insufficient intake can drastically affect bone health, but research suggests that it may also have other effects. Low-calcium diets can increase the accumulation of fat and weight gain; high-calcium diets can avoid fat accumulation. Epidemiological evidence and at least a related study on this showed that higher intakes of calcium minimize the chances of developing colon cancer [15]. Phosphorus has been identified as the second mineral that is maximum abundant in animals, and bones and teeth account for 80% of that chemical. Phosphorus functions in ossification processes with calcium and is a part of energy-rich compounds and nucleic acids. In the body, the levels of calcium and phosphorus are controlled by vitamin D3 and parathyroid hormones: parathormone and calcitonin [16]. Deficiency of phosphorus, which may be increased by low levels of vitamin D, leads to modifications in the human skeletal with the production of rickets in young animals. In most cases, vitamin D deficiency results from poor diet and insufficient sunlight exposure (UV radiation contributes to the transformation of vitamin D into active metabolites). Deficiency of phosphorus is caused by a low diet in this product, which distorts the Ca:P ratio (1.5–2:1), inhibits the absorption of Ca and P and their transport to the skeletal system [17].

2.2.2.4 Cobalt

Cobalt (Co) catalyzes biochemical reactions along various metabolic routes, including hydration, hydrogenation, and desulphurization. This factor participates in nucleic acid biosynthesis and stimulates the development of erythropoietin in the kidneys which is needed in living organisms for healthy erythropoiesis. Cobalt activates gluconolactonase and takes part in the pentose-phosphate pathway, an important tissue-oxidation process that is independent of the Krebs cycle. Glucose oxidation is most intensive in the mammary gland [18]. Half of the cobalt supplied to the rumen is used for the preparation of vitamin B12, which contains up to 4.5% cobalt and is also known as cobalamin. In ruminants, cobalt is primarily contained in the liver, kidneys and heart in limited quantities and its concentrations range from 3.0 to 220.0 $\mu g\ g^{-1}$ of fresh tissue weight [19]. Cobalt is necessary for the optimum production and function of ruminal microflora, and ruminants, particularly lambs, sheep, goats, calves, and adult cattle, are most prone to deficiency signs due to cobalt [20].

2.2.2.5 Copper

Within several enzymes, copper (Cu) is present and perform a significant function within biochemical activities. Copper is a part of two important enzymes of aerobic metabolism: superoxide dismutase and cytochrome C oxidase. The item discussed affects the prooxidative-antioxidant balance and exerts antioxidant effects. Copper is involved in the formation of myelin sheaths covering nerve fibers, hematopoiesis, and skeletal development; it controls collagen, elastin, melanin, and catecholamine synthesis; it affects hair coat color and waviness. Copper promotes uptake and metabolism of iron [20]. While deficiency symptoms due to copper are uncommon, the risk may be increased by other health conditions and other factors, which include copper metabolism genetic defects, absorption issues with high doses of zinc or vitamin C supplements, and other diseases.

2.2.2.6 *Zinc*

While deficiency symptoms due to copper are not common, the risk may be increased by other health conditions and other factors, which include copper metabolism genetic defects, absorption issues with high doses of zinc or vitamin C supplements, and other diseases [20]. Moreover, zinc has been identified as a crucial nutrient available in numerous natural food which might be consumed as a nutritional supplements. Zinc is available in numerous cold lozenges, and some medications marketed as cold remedies over the counter. Zinc participates in many facets of cellular metabolism. Approximately, 100 enzymes need it for catalytic activity and it is useful in immune function, synthesis of protein, healing of wounds, DNA synthesis, and cell division [12].

2.2.2.7 *Iron*

Iron is a mineral that occurs in most natural foods which could be applied as a nutritional supplement. They play a significant role in the proper functioning of the hormone, adequate performance of the cell, adequate functioning of the myoglobin and the fabrication of connective tissue [12]. They could exist in the marrow of the bone, spleen, and in the liver as hemosiderin or ferritin (a ferritin degradation product) or is in the muscle tissue in the myoglobin. Moreover, it has been observe that blood loss and losses of iron are higher in women that are on menstruation. The concentrations of hemoglobin are regularly calculated to find out whether there is enough iron in the body. Also, it has been stated that the final stage which entails iron shortage described as iron-deficiency anemia could happen when dose of hemoglobin available in the blood are available in minimal amount. The most widely used methods for screening patients for iron deficiency are hemoglobin and hematocrit tests, although they are neither sensitive nor precise [12].

2.2.2.8 *Sodium*

Food also contains sodium in large quantities, but without the addition of table salt (sodium chloride). In fact, the sodium retention mechanisms of the body are highly evolved and sodium shortages are uncommon, particularly for people on low sodium diets [2]. Sodium depletion can occur during persistent heavy sweating, diarrhea, or vomiting, or in renal disease [21]. Symptoms of low sodium in the blood (hyponatremia), include muscle fatigue, cramps, nausea, dizziness, and finally shock and coma. After intense heat exertion of high intensity, the sodium balance can be restored by drinking sodium and glucose-containing beverages (referred to as sports drinks) and by eating salted food. Drinking 1 L of water containing 2 ml (one-third teaspoon) of table salt should also fulfill one's needs [2].

2.2.2.9 Arsenic

It has been discovered that arsenic promotes the development of tissue cultures [22], and they have been recorded to play a greater function in the maintenance of health and efficiency of pigs and poultry. This function is close to that of antibiotics and seems largely related to the regulation of dangerous intestinal microorganisms. The fact that cadmium had been fractionated most especially those obtained from the horse kidney cortex in specific association with protein macromolecules indicates a biological function for cadmium [3].

2.2.2.10 Chlorine

In HCO_3-transporting erythrocytes, chlorine has been shown to be entailed electrolyte equilibrium and fluid maiantainace, gastric fluid, and alteration of chloride. Within extracellular fluid, chloride is the main anion. Moreover, it play a crucial role in the control of the osmotic pressure which occur extracellular and accounts for more than 60% of the anions in this fluid compartment [23]. It is the gastric juice's main anion and is surrounded by almost identical quantities of hydrogen ions. The chloride in the gastric secretions is resulting mainly from blood chloride and is usually assimilated in the lower intestine during the latter stages of digestion [3].

2.2.2.11 Iodine

Iodine deficiency disorders are the most frequent cause of preventable brain damage that affects about 50 million people worldwide. Severe iodine deficiency can result to the impairment of the developing fetus during pregnancy; this can bring about cretinism (a mental retardation that cannot be reversed with developmental defects and short stature) as well as miscarriage or death [24]. Certain more prevalent impacts due to deficiency in iodine include decrease in cognitive and neuromuscular functioning. Ocean is a good source of iodine, but the amount iodine in food changes more than coastal areas and generally represents the amount in soil. During chronic iodine deficiency, the thyroid gland extends from the blood for the production of thyroid hormones because of its attempts to collect more and more iodide (a process in which the iodine acts in the body); it gradually grows into a conspicuous lump at the front of the neck, a disorder known as a "goiter" [2]. Many foods, such as sweet potatoes, cassava, some beans, millet, and others in the family of cabbage, have a substance known as goitrogens that interfere with synthesis of the thyroid hormone. While destroyed by cooking, the substances may be an important factor in people with the already existing iodine deficiency who rely on goitrogenic foods as staples [25].

2.2.2.12 Potassium

Potassium has been recognized as element that performs a vital role in the balance of the acid-base balance in the regulation of Na^+/K^+-ATPase, maintenance of the cell membrane, osmotic pressure, as well as the contraction the cardiac muscle. Moreover, they also involved in several enzymatic cellular reactions [23]. Aldosterone regulates its metabolism. Hyperkalemia of serum potassium is elevated, which might led to the development Addison's disease and unconventional chronic renal disaster as well as the development of hypokalemia as a result of low serum potassium which might lead to metabolic alkalosis and diarrhea and which might exist in the periodical paralysis in the body. Also, it has been observed that the process involved in the milk production is decreased markedly when lactating dairy cows have hypokalemia. Deficiency diagnosis or symptoms arise due to illness, functional and physiological disorders involving impaired skeletal, smooth, and cardiac neuromuscular functions, muscle fatigue, paralysis, and mental distress [26]. Others include cardiac arrythmias, reduced resistance to glucose, and transformed electrocardiogram in calves. Moreover, it has been observed that Potassium shortage might have adverse effect on the accumulation of kidney tubules, contributing to the failure to concentrate urine, which also induces improvements in stomach secretions which intestinal motility [3, 27]. Twelve elements are defined as indispensable for good health. Calcium, P, Mg, and F, for example, are important for bone and membrane structural functions. Mineral dust exposure can result in a large variety of breathing problems. Dust may be produced from mining rocks or coal, sandblasting, and smoking feathers from fires (both natural and man-made) or simply from the winds that disperse fine-grained minerals from the earth's surface [28].

2.2.2.13 Fluoride

Fluoride contributes to the deposition of minerals in bones and teeth, which protects against decay of the tooth. The US epidemiological research of the 1930s and 1940s showed an inverse association between dental caries and natural fluoride in water levels [2]. In areas where fluoride levels are low in drinking water, fluoride supplements are prescribed for children over the age of 6 months; dentists may also periodically apply fluoride rinses or gel to their patients' teeth [2, 21, 25]. Fluoridated toothpastes are effective sources of fluoride for adults and also children, who tend to benefit from fluoride ingestion [24]. Like calcium, magnesium helps to build bones and teeth. It also helps control blood pressure and blood sugar and allows muscles to contract, nerves to transmit signals, blood to clot and enzymes to work.

2.2.3 Application of Geochemistry on Human Health

Role of geo-chemistry for the health of human has been identified to perform a crucial function in the maintenance of human health. Number of trace elements present in the soil which shows effectiveness in different ways includes fluorine, lead, iodine, selenium, zinc, thallium, aluminum, copper, cadmium, and arsenic. The amount of this element in soil depends on the soil fertility and human action like pollution. If the quantity of trace element in soil increase from average or decrease, it could causes problem related to health, e.g., due to deficiency of selenium a health-related issue which occur is Kashan disease, and if their level increase then itai-itai problem could occurs which is a chronic disease associated with a very high level of cadmium.

Many researchers have done their research for this purpose but many more advancement is still required in this field [29]. When we discuss the intake of fluorine than high level of this element increased the chances of dental caries, some other elements like Pb, Mo, and selenium shows protective effect for the health of human. Cardiac diseases has been identified as a one of the important health related issue which could be linked to geo-chemical environment, the chances of this disease mostly increase in those individuals who are used to smoking and some other factor like hypertension and dietary also important. Some factors like temperature, rainfall, and hardness of water affect the geochemical nature. The importance of Arsenic has been established in different fields like medicinal, industrial, and agriculture but in human it also shows some toxic effect when the level of this element increases above average level. Also, it could show more adverse effect on human skin as well as affects the immune system after the exposure of the body to this element.

Iodine is an important trace element which has useful effect for human health and if the level of these element increases or decrease than their average levels then it could result into a typical problem for human. Also, sea-food is an abundant source for this element but their presence in a very higher amount could lead to a very high toxicity. A typical example of such trace element includes uranium, cadmium, arsenic, mercury, and lead [30].

Radwińska *et al.* [13] discussed in their research the effects of low levels of macronutrients and micronutrients as essential factors that are essential to improve the growth of lambs, calves, and children. Rim *et al.* [14] proposed that toxicological reports on some compounds in the area of occupational health. This was based on that data obtained and the inquiry into toxicological evaluations of rare earths and their safety impacts performed on their employees. Renwick [15] reviewed micronutrient toxicology and evaluated their adverse effects. The authors concluded that the normal risk evaluation protocols for poisonous substances available in food. Sudhir *et al.* [31] in their work on their work on micro minerals observed their great influence most especially in improving the rate of reproduction and preventing numerous reproduction challenges in animal.

2.3 Conclusion and Recommendations

Therefore, this chapter has provided comprehensive information on the application of these essential mineral for the treatment of human diseases most especially the role of geo-chemistry for the restoration and maintenance of human health. It was also established that geo-chemistry played numerous role in human health through the utilization of these essential trace dietary element for the management of several deficiencies, diseases disorders, tropical illness, and nutritional deficiencies. Therefore, there is a need to perform the structural elucidation of major essential trace dietary element that have not been fully characterized. The application of novel techniques such as nanotechnology, cheminformatics, informatics, bioinformatics, and structural biology, as well as the utilization of internet of all things and artificial neural network will play crucial role in the prediction of active component present in raw material and their eventual application for the management of diseases and disorderliness.

References

1. Stephen, N.C., Alexander, L.U., Anthonia, O.N., Uchebike, O.G., Application of Environmental Geochemistry Research to Public Health Issues in Nigeria. *Am. J. Environ. Prot.* Special Issue: Integrating Earth Materials, Diet, Water and Human Health, 3, 6–2, 30–34, 2014.
2. Weininger, J., *Nutritional disease*, Encyclopaedia Britannica, Chicago, USA, https://www.britannica.com/science/nutritional-disease (2020, November 11), 2019.
3. Soetan, Olaiya, C.O., Oyewole, O.E., The importance of mineral elements for humans, domestic animals and plants: A review K. O. *Afr. J. Food Sci.*, 4, 5, 200–222, May 2010.
4. WHO, *World Health Organization model list of essential medicines: 21st list 2019*, World Health Organization, Geneva, 2019. https://www.who.int/groups/expert-committee-on-selection-and-use-of-essential-medicines/essential-medicines-lists, 2019.
5. UNICEF, *Vitamin A Deficiency and Supplementation UNICEF Data*, https://data.unicef.org/topic/nutrition/vitamin-a-deficiency/, 2018.
6. Frank, D., Galloway, J., Assmus, K., *The Life Cycle of a Mineral Deposit*, Denver Federal Center, Denver, https://pubs.usgs.gov/gip/2005/17/, 2005.
7. DeVivo, B., Belkin, H., Lima, A., *Environmental geochemistry: site characterization, data analysis and case histories*, Elsevier Science Publisher, Edinburgh, United Kingdom, 2017.
8. Harris, N.B., Pearce, J.A. *et al.*, Geochemical characteristics of collision-zone magmatism. *Geol. Soc.*, London, Special Publications, 19, 1, 67–81, 1986.
9. Reimann, C. and de Caritat, P., Distinguishing between natural and anthropogenic sources for elements in the environment: regional geochemical surveys versus enrichment factors. *Sci. Total Environ.*, 337, 1-3, 91–107, 2005.
10. Madejón, E., De Mora, A.P. *et al.*, Soil amendments reduce trace element solubility in a contaminated soil and allow regrowth of natural vegetation. *Environ. Pollut.*, 139, 1, 40–52, 2006.
11. Thornton, I., Environmental geochemistry and health in the 1990s: a global perspective. *Appl. Geochem.*, 8, 203–210, 1993.
12. Merrell, P.H., Technical Market Manager, Jost Chemical Co. PhD 2020, *The Importance of Minerals in the Long Term Health of Humans*, 2020, Technical Market Manager, Jost Chemical Co. 2015, 2020, PubMed.
13. Radwińska, J. and Żarczyńska, K., Effects of mineral deficiency on the health of young ruminants. *J. Elem.*, 19, 3, 915–928, 2014.
14. Rim, K.T., Koo, K.H., Park J.S., Toxicological Evaluations of Rare Earths and Their Health Impacts to Workers: A Literature Review. *Saf. Health Work*, 4, 1, 12–26, https://doi.org/10.5491/SHAW.2013.4.1.12, 2013.
15. Renwick, A.G., Toxicology of Micronutrients: Adverse Effects and Uncertainty. *J. Nutr.*, 136, 2, 493–501, 2006.

16. Sobiech, P., Rypuła, K., Kotwica, B., Michalski, S., Usefulness of calcium-magnesium products in parturientparesis in HF cows. *J. Elem.*, 15, 4, 693–704, 2010.
17. Iqbal, M.U., Bilal, Q., Muhammad, G., Sajid, M.S., Absorption, availability, metabolism and nexcretion of phosphorus in ruminants. *Int. J. Agric. Biol.*, 7, 689–693, 2005.
18. Kennedy, D.G., Kennedy, W.J., Blanchflower, J.M., Scott, D.G., Weir, A., Malloy, M., Young, P.B., Cobalt-vitamin B-12 deficiency causes accumulation of odd-numbered, branched-chain fatty acids in the tissues of sheep. *Br. J. Nutr.*, 71, 67–76, 1994.
19. Sivertsen, T. and Plassen, C., Hepatic cobalt and copper levels in lambs in Norway. *Acta Vet. Scand.*, 45, 2, 69–77, 2004.
20. Hostetler, C.E., Kincaid, R.L., Mirando, M.A., The role of essential trace elements in embryonic and fetal development in livestock. *Vet. J.*, 166, 125–139, 2003.
21. Westport, C.T., *Nutrition in first 1,000 days: the state of the world's mothers 2012*, p. 16, Save the Children, Westport, CT, 2012.
22. Hayes, D., Adverse effects of nutritional inadequacy and excess: a hormetic model1–4. *Am. J. Clin. Nutr.*, 88, suppl, 578S–81S, 2008.
23. Murray, R.K., Granner, D.K., Mayes, P.A., Rodwell, V.W., *Harper's Biochemistry*, 25th Edition, McGraw-Hill, Health Profession Division, USA, 2000.
24. Tulchinsky, T.H., Micronutrient Deficiency Conditions: Global Health Issues. *Public Health Rev.*, 32, 243–255, https://doi.org/10.1007/BF03391600, 2010.
25. Young, E.M., *Food and development*, p. 38, Routledge, Abingdon, Oxon, 2012.
26. Balmurugan, B., Ramamoorthy, M., Shankar, R., Katiyar, R., Mineral an Important Nutrient For Efficient Reproductive Health In Dairy Cattle. *Int. J. Sci. Environ. Technol.*, 6, 1, 694–701, 2017.
27. Godswill, A., Health Benefits of Micronutrients (Vitamins and Minerals) and their Associated Deficiency Diseases: A Systematic Review. *Int. J. Food Sci.*, 3, 1, 2520–46371–32, 2020, www.iprjb.org1.
28. Plumlee, G.S. and Morman, S.A., Mine wastes and human health. *Elements*, 7, 399–404, 2011.
29. Oliver, M., Soil and human health: a review. *Eur. J. Soil Sci.*, 48, 4, 573–592, 1997.
30. Fuge, R. and Johnson, C.C., Iodine and human health, the role of environmental geochemistry and diet, a review. *Appl. Geochem.*, 63, 282–302, 2015.
31. Sudhir, P.S., Upadhyay, S.K., Pandey, A., Kumar, S., *Molecular Approaches in Plant Biology and Environmental Challenges*, Springer Nature Singapore Pte Ltd., 2011.

3

Applications of Geochemistry in Livestock: Health and Nutritional Perspective

Charles Oluwaseun Adetunji[1]*, J. Inobeme[2], Inamuddin[7], Muhammad Akram[3], A. Inobeme[5], Khuram Shahzad[3], Maliki Munirat[5], Saher Islam[8], Noshiza Majeed[4] and S.O. Okonkwo[6]

[1]Applied Microbiology, Biotechnology and Nanotechnology Laboratory, Department of Microbiology, Edo University Iyamho, Auchi, Edo State, Nigeria
[2]Department of Geography, Ahmadu Bello University, Zaria, Nigeria
[3]Department of Eastern Medicine, Government College University Faisalabad, Pakistan
[4]Minhaj University Lahore, Pakistan
[5]Department of Chemistry, Edo University Iyamho, Iyamho, Nigeria
[6]Department of Chemistry, Osaka Kyoiku University, Osaka, Japan
[7]Department of Chemistry, King Abdulaziz University, Jeddah, Saudi Arabia
[8]Institute of Biochemistry and Biotechnology, Faculty of Biosciences, University of Veterinary and Animal Sciences, Lahore, Pakistan

Abstract

Geochemistry is the branch of science that uses the tools and philosophies of chemistry to give detail of the mechanisms regulating the working past and present of the earth. Geochemistry has an application in livestock like geochemistry in animal health and geochemistry and the nutrition of agriculture livestock. Geochemistry has been identified with several benefits most especially in animal health which involves dealing with the pathological and metabolic effects of different elements on animal health as well as in the nutrition of agriculture livestock which deals with the amount of ingested elements by the grazing animals. Therefore, this chapter intends to highlight the benefits of geochemistry in livestock most especially in the adequate maintenance of animal health and animal nutrition. Special highlights were also made on the significance of numerous minerals that have several roles in effective production of livestock.

**Corresponding author*: charliguitar@yahoo.com; adetunjicharles@gmail.com; adetunji.charles@edouniversity.edu.ng

Inamuddin, Mohd Imran Ahamed, Rajender Boddula and Tariq Altalhi (eds.) Geochemistry: Concepts and Applications, (37–56) © 2021 Scrivener Publishing LLC

***Keywords*:** Geochemistry, animal health, livestock, nutrition, animal diseases, minerals

3.1 Introduction

Geochemistry exerts a fundamental impact on the concentrations of mineral nutrients in the soils on which we cultivate the crops and animals that form the human food chain. Owing to the various geological environments, the mineral status of humans, plants, and crops varies considerably across the globe. Since diet is the most important source of mineral nutrients in livestock, understanding the biogeochemical controls on environmental mineral distribution and mobility is crucial to assessing toxicity and health risks [1]. Adequate intake and absorption of trace minerals is needed for a range of metabolic tasks including responses by the immune system to pathogenic challenges, reproduction, and development. Mineral supplementation approaches are increasingly becoming complex, as variations in trace mineral content of both avian species and livestock are critical to achieving optimum performance in current animal production systems.

Subclinical or marginal deficiencies can pose a greater problem than acute mineral deficiencies because clear clinical symptoms are not obvious to enable the manufacturer to identify the deficiency; however, animals continue to develop and reproduce but at a reduced rate [2]. The adverse consequences of the production of livestock on the environment are becoming more and more severe at all levels—local, regional, national, and global—and needs to be addressed fast. According to the FAO, when land-use emissions are taken into account, the livestock generally releases carbon dioxide emissions of about 9% from human activities and methane emissions which could be estimated about 37%. Gas primarily from cattle's digestive system and other domesticated ruminants—and 65% of nitrous oxide pollutants, particularly from manure. The issues affecting livestock production cannot be considered alone nor are they limited to the effects on the environment [3]. There is a need to pay attention to the occurrence of livestock living closer to the house in many regions worldwide raises concerns about the consequences for public health. There has hardly been any research following the rapid growth of livestock farms that of large-scale, continually blended with developed areas, and their possible effects on adjoining resident safety. Numerous crowded farms with livestock are found in a place where there is a big population in many regions around the world, most especially in the Western world and Asia [4].

So, what agricultural contaminants are of possible public health relevance? Based on local conditions, methods of manure disposal, and environmental laws, the water and quality of the soil can be affected. Livestock can present a significant toxicological risk to human health, both indirectly and directly, depending on the nature of the human-animal relationship [5]. Adequate knowledge about the function of inorganic constituents most especially in the preventions of epidemiology of clinical symptoms found in animals has pointed put the significant of geochemistry [4]. Equally complex are possible health consequences from farm pollution, which include zoonotic diseases, antimicrobial resistant as a consequence of infections due to bacterial and respiratory illnesses. In the One Health framework, the combination of different discipline and integrated method to tackle possible or current risks occurring at the boundary between human and animal environments [6]. Therefore, this chapter has provided detailed information on the significant of geochemical in the prevention of several diseases facing livestock and their critical role when included in the nutrition of animals. Special highlights were also made on the significant of numerous mineral that have several roles in effective production of livestock.

3.2 General and Global Perspective About Geochemistry in Livestock

The nutritional deficiencies and nutritional excesses in grazing livestock are related to geochemical nature of soil contents. This has been acknowledged since 1960s and several works has been carried out in Ireland, Soviet Union, Australia, United Kingdom, Ireland, New Zealand and USA, respectively [7]. Their major researches have focused mostly on lack of Cu, Se, and Co in sheep and cattle while special attention was placed on the role of S and Mo in soil nature. The contact of heavy metals in those areas contaminated with smelting activities and mining gain attention have also been validated [8]. The essential trace elements and toxic metals both are influenced by soil and plant factors including their bioavailability have been studied [9]. The soil has been investigated as major contact pathway for cattle feeding on graze that have been exposed to heavy metal [10]. Moreover, there is a close relationship between trace minerals concentration in rocks, soils derived from them, drainage sediments, and trace minerals deficiency issues in cattle [1]. Geochemical mapping indicates the concentration of elements in the soil environment and give information about their dispersion and distribution [11]. For establishing the levels of trace minerals in soil, environment regional geochemical mapping has been identified in this regard [12]. Also, some geochemical factors governing different elements levels in drainage residues such as climate, dispersion, chemical mobility, topography, and metals dilution have been highlighted during the process of transport [13]. For the detection of metalloids minerals concentrations geochemical mapping has been utilized for this purpose [14].

Minerals imbalances or deficiencies in fodders are accountable for reproduction problems and stumpy production amongst grazing livestock in developing nations [15]. Mineral deficiencies and excesses have been investigated in developing countries in tropics regions of world [16]. Cu deficiency has been investigated in Japan, New South Wales, Tasmania, Gulf Emirates; Co deficiency in Australia; Se deficiency in Egypt, Gulf states, Israel, Libya, and Zimbabwe [17–19]. In developing countries grazing livestock totally depend on local forage for maintaining their mineral requirements [20, 21]. The most vital factors governing trace minerals availability and concentrations in plants are soil pH, soil temperature, soil moisture, drainage conditions and seasonal variations, fertilization, and plant species [22]. Soil forming processes, weathering, and chemical speciation within the soil control the release of trace

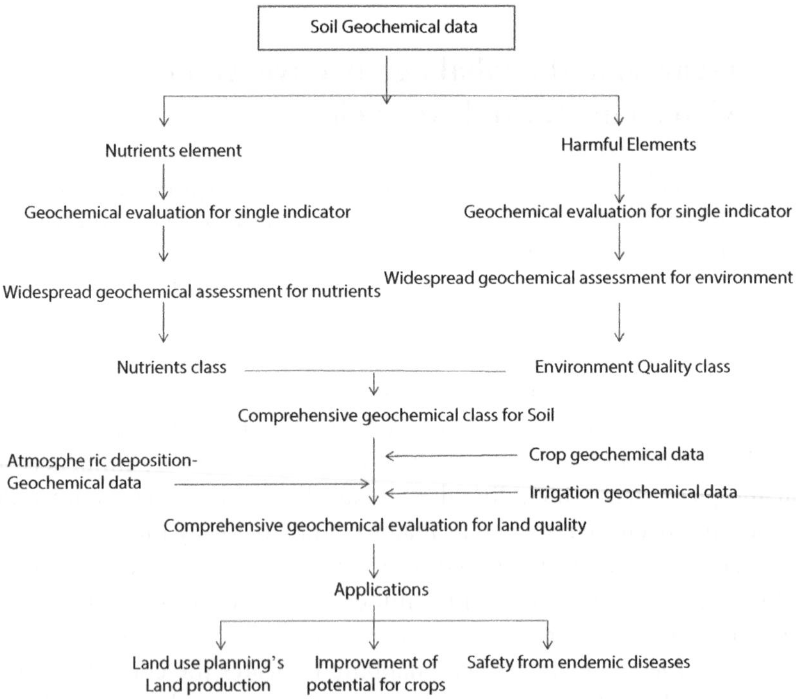

Figure 3.1 Geochemical evaluation of land for better crop production for livestock.

elements to plants [23]. The irrigation, agricultural product safety, air and water quality in combination with soil nutrients, and environmental quality articulate geochemical class and land quality [24] as shown in Figure 3.1.

3.3 Types of Geochemistry and Their Numerous Benefits

Geochemistry is the branch of science that uses the tools and philosophies of chemistry to give detail of the mechanisms regulating the working past and present of the earth [25]. It is wide and interesting subject both in terms of topics addresses and techniques employed. Geochemistry part of the earth systems and environmental science contain number of articles debating state of the art information on the techniques and technical advances available to the recent geochemist and in what way the tools and principles of the chemistry have been carried to tolerate on the full range of problems in earth and environmental sciences [26]. The geochemistry is divided into various types and these types include:

- Analytical geochemistry
- Isotope geochemistry
- Low temperature geochemistry
- Organic and petroleum geochemistry
- Cosmo chemistry
- Environmental geochemistry

Analytical geochemistry: Although the number of things can be learnt by the application of theoretical calculations and estimates within geochemistry, the number of geochemical studies falls within the kingdom of analytical geochemistry and are apprehensive in a way with the measurement of samples. Geochemical analysis concerned with the measurement of quantities of isotopes, elements, and minerals in the rock, natural liquid, and in the naturally occurring gas. Isotope geochemistry: This branch concerned with the relative and absolute concentration of the constituent and there isotopes in section from the soil and solar system. Low temperature geochemistry: This branch mainly concerned with the chemical processes happening on the earth crust. Organic and petroleum geochemistry: This branch is mainly concerned with the biogeochemical cycles with the carbon cycles probably being one of the most important devoid of which life could not exist. Cosmo chemistry: This is apprehensive with the scientific questions being spoken. Environmental geochemistry: It is concerned with the chemical composition of rock, water, microorganism, air, soil, and organisms that exist on the earth crust [27].

Geochemistry has number of applications in different fields; here, we will discuss the application of geochemistry in livestock (geochemistry and animal health and livestock nutrition). Geochemistry and animal health: Disease and less reproduction due to imbalance in inorganic components this problems mainly associated with the deficiencies of Co, Cu, I, Zn, Na, P, and Se in farm livestock and the effects including pathological and metabolic pathological effect due to chronic exposure of animal to toxic components and metabolic effects are mainly due to deficiency of these above mentioned constituents. Geochemistry and nutrition of agriculture livestock: In 1960, 1970, and 1980, mostly concentration was on the nutritive requirement of native livestock and inside this period geochemical map leveraging on the available sample and investigation of earth to determine the areas of nutritional deficiency and elements excess [28].

Geochemistry is the branch of the earth science that uses the instruments and principles of chemistry to enlighten the mechanism regulating the working past and present of the earth. The term of geochemistry is wide and spread to the study of the formation of earth and solar system and also covers the topic of ocean, earth crust and atmosphere. It is a wide and interesting subject both in relations of issues it addresses and the techniques employed. Victor Goldschmidt (1888–1947) is the father of geochemistry; one of his accomplishments was to begin to categorize the behavior of the chemical components in an earth science framework and its classification according to their favored host levels on silicate earth (lithophile, siderophile, chalcophile, and atmophile). Thus, the wide subjects like geochemistry has naturally subdivided into different types these different types are different from each other and in number of different cases they are overlap to each other and these types are as under:

- Analytical geochemistry
- Isotope geochemistry
- Low temperature geochemistry
- Organic and petroleum geochemistry
- Cosmo chemistry
- Environmental geochemistry

3.3.1 Analytical Geochemistry

Although abundant can be educated by the applications of theoretical calculations and estimates inside geochemistry but the majority of geochemical studies are concerned with the quantities of samples. Geochemistry involves in the measurement of quantities of element, isotopes and minerals in a soil or naturally present gases. Different areas like isotope geochemistry are ruled by techniques such as mass spectroscopy while with different varieties of sample being analyzed and specific system of the machines; in a field like sedimentary, igneous, and metamorphic geochemistry, or some other branches of geochemistry closely related to petrology, a number of analytical techniques are used depending upon either the bulk rock configuration or glass chemistry and minerals are of interest, mineral analysis including the determination of chemical relation between the mineral grains, microanalytical techniques are important such as x-rays, and laser procedures are most commonly used for the analysis of major and trace elements and it made a great contribution in the development of modern theories of ore genesis and diagenesis [29].

3.3.2 Isotope Geochemistry

This branch is mainly concerned with the relative and absolute concentration of element and their isotopes in testers from the soil and solar system. This type of measurements offers powerful instruments to investigate a variety of scientific problems from the planet origin to past climate change and the source of elements in a number of geological reservoirs so it can be divided into two types:

- Radiogenic isotopes
- Stable isotopes

So, the general instrument of the isotopes geochemistry is uses in a number of ways like the use of known decay rates of the radiogenic isotopes within the geochronological techniques is well known and common in the earth and environmental sciences for checking the age of tester or the rate of process. Stable isotopes are also have fundamental influences to understand the process such as the rock and water cycles and changes in the earth system over the geological time [30–32].

3.3.3 Low Temperature Geochemistry

It deals with the chemical process occurring under the conditions established at or at the surface of the earth's crust. It is in comparison with the igneous geochemistry where processes mostly occur at greatly upper temperature and pressures. Its important type includes environmental geochemistry [33].

3.3.4 Organic and Petroleum Geochemistry

This type deals with the dissemination, conformation, and outcome of organic matter in the geosphere in both majority and molecular level on the combination of the features of geology biology and chemistry it mainly deals with biogeochemical cycles like with carbon cycle which is one of the most important and without it life cold not possible, it gives the basis of fossil fuel like crude oil, natural gas, and coal through the fusion into deposits and burial of biomass from decayed organism. Organic chemistry is closely linked with petroleum chemistry [34].

3.4 Application of Geochemistry in Livestock

Geochemistry has number of applications in different fields; here, we will discuss the application of geochemistry in livestock, and we will discuss about geochemistry and animal health and geochemistry and the nourishment of agriculture livestock [7].

3.5 Geochemistry and Animal Health

Disease and less reproduction due to imbalance in the level of inorganic elements cause substantial economic loss to British animal industry. Pathological and metabolic defects most commonly associated with decrease level of Cu, Co, I, Se, P, Na, and Zn elements pathological signs may appear earlier the evident signs of deficiency appear. Appetite, growth resistance to infection, and reproduction may be compromised earlier the peripheral signs of deficiency appear. Pathological effects occur due to long term contact of animal to the harmful elements like Cd, Fe, F, Pb, Se, and Zn the extensive range of assessed important elements required in high dose at the time of high metabolic activity like rapid growth and effects on the immersion of element from the digestive track and causes changing response of element to their geochemical environment compromise [35].

- Inorganic composition of dietary constituents.
- Physiological variable including the confiscation of elements in animal tissue and alteration of animal to this imbalance of elements.
- The consequence of inorganic inequality on the availability of elements in the diet and on acceptance of the animal population.

Supplements are rich in these elements that may tackle the deficiencies that may come from geochemical conditions the time upon which symptoms seem due to the lack of diet that is affected by the surviving of tissue reserve of elements these diverge between animals and postponed the metabolic effects to develop. Agriculture activity was related with increased level of calcium and magnesium, alkalinity, and also potassium and sodium city areas were related with increased concentration of sodium, potassium, and calcium other factors were not considered in sampling event and as well as not in context of biogeochemical sources and processes when all data are mixed then we will find that there is a relationship among wetlands and stream nitrate concentration. The barren lands are also related with increased concentration of sulfate levels and upon the combination of data collected over 2 years during three synoptic sampling event appear to increase our ability to recognize the processes that connected the land use and stream chemistry.

3.6 General Overview of Geochemistry in Livestock's Merits of Geochemistry/Essential Minerals in Livestocks

Minerals have an enormous impact on the reproductive physiology of animals, and their imbalance causes numerous problems which lead to reduction in the efficiency of production. Most notably, dairy animals struggle with nutritional deficiencies due to high demand and inadequate feeding, resulting in poor success in reproduction. Minerals are structural components of the body and play a vital function as components of body fluids and tissues and as regulators of cell replication and differentiation in the actions of enzymes and hormones. Minerals are essential to all animal physiological processes as well as reproduction [36].

Deficiency in copper is linked with high molyb-biological processes, such as metallo-denum levels component as crops cultivated in "test soils" (having co-factors and enzymes). Such plays have increased Mo and low Cu, both as high organic matter. Mo activator of enzymes involved in the free radicals intracellular detoxification process, and in and Cu interactions further lower the Cu available for absorption, secondary molecules stabilization. The important sign related to reproduction in is component of hormones and thus directly declines in fertility in cattle. Steroidal changes control endocrine function. Metabolism can result in altered reproductive behavior: metabolism of carbohydrates and proteins and nucleic acids, such as nymphomania in the ewe [37]. Any alteration in its level can alter the development of Copper alongside with deficiency of chromium in reproductive and other hormones delayed onset [38]. The contemporary animal production system strategies face a challenge in handling the mineral nutrition. Mineral nutrients availability is a major problem facing livestock and this also play a vital role in affecting the physiological balance in animals [39].

It has been identified that trace element plays a crucial role during the growth of the animal as well as during the reproduction [7]. Moreover, these trace elements have been identified to play a crucial role toward the development of animal feed. The shortage of this trace element might lead to high level of deficiency in the animal [40]. Conventional farming has various resources to resolve trace mineral deficiencies in trace minerals and/or imbalances, such as offering animals complementary winter feeding, increasing pasture quality through the use of pesticides and fertilizers, or pharmacological management of infections with parasites, interventions that could improve the nutritious level of the animal through adequate administration of essential mineral to the animal [41]. Minerals are inorganic elements present in the body, in limited quantities. Inorganic means carbon is not in the material. Minerals are included in livestock feed rations to avoid mineral shortages and are made available by free access to mineral and salt blocks. Mineral deficiencies can lead to low weight gain, poor feed efficiency, and poor reproductive characteristics. Minerals are known as macrominerals [37] or microminerals. For repair and growth of bone, alongside other body functions, phosphorus and calcium are required in a certain ratio. Magnesium is required for skeletal growth and for biochemical reactions in muscles. Grass tetany, a potentially deadly disease that occurs on spring pastures while cows are lactating, is due to magnesium deficiency.

Potassium enhances the absorption of glucose. Chlorine and sodium are provided by salt (NaCl) and are required for the maintenance of water balance and to provide dietary iodine. Sulfur is needed for synthesis of protein (EdLibrary.com). Adjusted weaning weights in calves have been shown to improve trace mineral supplementation to the cow/calf pairs. Summary results of 22 feedlot trials showed a substantial increase of 3.26% in daily benefit and a boost of 4.05% in feed conversion when zinc methionine was applied to the control diet. The trials were performed under varying environmental conditions at different locations and using variable cattle sources. Macro minerals and trace elements are equally necessary as they participate in body structural component development and proper enzyme, hormone, vitamin, and cell functioning. Mineral deficiency and toxicity are unique to the environment concerned. Minerals have beneficial or adverse effects on the physiological health of animals, depending on their balance [37]. Besides energy and protein, mineral also plays a significant role in animal development and reproduction. Deficiencies of some mineral elements can cause reproductive disorders because minerals play vital role in livestock health and reproduction. A significant issue in our livestock population has arisen reproductive disorder and lack of fertility. Adequate mineral supplementation depends upon age, pregnancy period, and lactation. Macrominerals and trace minerals are all important as they form structural components of the body and work properly with enzymes, hormones, vitamins, and cells [37].

3.6.1 Specific Examples of Authors That Have Used Essential Minerals in Livestock

Thornton [7] focused on mineral nutrition in wildlife and farm animals and emphasis on the application of geochemical maps for maximum land utilization for effective wildlife protection as well as their breeding. In his research on selenium geochemistry and health, Fordyce [1] concluded that the Se status of humans, animals, and crops varies considerably across the world due to various geological conditions. Since diet is the most significant source of Se in humans, understanding the biogeochemical controls on environmental distribution and mobility is crucial to assessing their health status [42]. Meyer *et al.* [4] reported on the effect of geochemistry on animal and human health hazards in geographically distributed livestock production systems which result to the high likelihood of multiple consumption routes of hazardous inorganic ingredients for livestock and humans was recommended.

Kwon *et al.* [5] studied the impacts of pig burial leachate and heap of cow manure on the microbiological and geochemical features most especial from water samples from agriculture, which entails leachate, well monitoring water derived from stream, groundwater from well. Their study highlights the need for a deeper understanding of microbial population dynamics available around the groundwater which play a crucial role in effective determination of the level of contamination in these water bodies. Thornton [7] in their work on geochemistry and mineral nutrition of wildlife and agricultural livestock opportunities for future research, applied geochemical maps for optimizing land use, productive livestock development, and enhancing wildlife protection. In their research, soil geochemical details obtained from sample sites in counties that documented anthrax in livestock and wildlife since 2000. Balamurugan *et al.* [37] stated that macro minerals and trace minerals are equally significant because they are involved in the development of body structural components and the proper functioning of enzymes, hormones, vitamins and cells.

3.6.2 Livestock in Relation to Geominerals

The demanding growth in livestock sector forces many ecosystems and take part in creating environmental problems globally. Because livestock are provided with substantial amount of minerals and are evacuated into feces and urine, and their concentrations increase as the mineral quantity over mineral necessities increase [41]. Hence, heavy minerals foods contaminate the environment and this wrong policy of mineral diet leaves behind mineral footprint on relevant form [39]. The future of livestock exists in balancing the two challenging loads that are animal food products and environmental services [43]. In the same way, heavy mineral supplementation following enriched mineral foods creating risk for environment as well as consumers because of increasing toxic metal residues [39]. The livestock sector is playing a key role in agriculture economy. The animal source products demand will be more than other food items because of increasing populations and incomes, it is estimated that the meat demand is more increasing from year to year [44]. In current years, the public concern has been increased about the environment damage initiated by excessive animal feeding processes [43]. The FAO and other European Community have established mineral concentrations levels in foodstuffs for protecting livestock, the consumers, and the environment [39]. Possibly, the environmental concerned species are poultry, cattle, and pigs and the major main minerals are Zn and Cu [39]. Hence, Cu and Zn are extensively used growth stimulators in industries [45]. In pigs, dietary concentrations of copper sulfate and Zn sulfate is 150–250 mg/kg and 2,500–3000 mg/kg, respectively, have been revealed to boost growth without any poisoning risk to animals [39].

3.6.3 Trace Minerals Parallel Importance in Livestock

For optimum regulation of all biochemical processes, trace minerals are required for numerous biological developments, which is also essential for the maintenance of animals' health and productivity [46]. Optimum nutrition along with suitable trace minerals assures catalytic, physiological, structural, and regulatory functions in organisms [47]. It is established that the trace minerals must be delivered to livestock in optimum levels in accordance with needs during swift growth and development of animals and reproduction cycle [40]. The different authorities like INRA in

France, FEDNA in Spain, and ACR in the United Kingdom have suggested mineral requirements to guarantee that livestock production not disturbed by dietary imbalances [39]. The USA recommendations for livestock are imaginably more up to date [48]. Some minerals are antagonists to each other, so minerals recommendations include safety margins. Such as, in ruminants, Mo, So, and, to some extent, Fe prevent the uptake of Cu and Zn uptake is inhibited by high levels of Ca. It is also recognized that in the presence of Zn, higher concentration of Cu is required and, in case of stress animals, requires greater Cu and Zn levels [49].

Furthermore, when defining mineral recommendations, the most important factors like raw materials type, quantity, storage condition, diet processing, environment condition, and contents of other minerals must be considered [50]. Although, the importance of trace elements in human and animal health is sound recognized but there are various overlooked nutrients present in animal's diet having physiological role but undervalued [41]. These are necessary for optimum growth and reproduction, for maintaining body functions and stimulating immune responses for determining health status [48]. Certainly, the symptoms of mineral deficiency and unbalances may not always be apparent; sometimes, it might be very difficult to recognize the influence of inadequate trace elements [51]. Some trace minerals have been highlighted to play numerous roles. For instance, Se has been renowned for a substantial time to be essential for fertility and growth in animals and helpful in preventing various diseases, recently recognized as integral part of various enzymes, functioning as antioxidants [52].

Similarly, numerous trace minerals may perform a solitary function, like Cu, Se, and Zn are all playing role in immune function, in this way rather making an allowance for the individual function of trace minerals, it is essential to reflect the function of trace minerals in certain common problems as a whole come across in livestock [50]. In various areas where mineral deficiencies and mineral toxicities were a tricky in animals globally and identified and managed by dietary supplements, by that period, animal production was directed by traditional way, pretty related to soils and locally made feed, this might be linked to the high level of mineral imbalances. The request for livestock meat production has risen drastically due to the fact that world's population including developing countries increases day by day [53].

3.6.4 Heavy Metals Impact Livestock

The higher concentrations of heavy metals contents of mineral enhancements in comparison to feedstuffs and relations with other important minerals can cause noxious influence on animal's health [49, 54, 55]. Actually, mineral supplements deliberated as one of the feedstuffs along greater poisonous heavy metal dose and highest allowable dose are higher than other feed stuff established by European Union [39].

Cu is essential as a cofactor for many life processes, almost all organisms have established extremely dedicated homeostatic tools to recruit, eliminate Cu and neutralize its toxic effects because its excess is extremely toxic [56]. There are noticeable distinctions of tolerance in domestic animals to distended levels of nutritional copper [57]. Among the domestic animals, sheep are the utmost vulnerable to toxicity of Cu and its supplementation limited to 15 mg/kg DM [39]. It has been confirmed that sheep when compared to pigs have narrow capability to accrue Cu joined to metallothionein in the livers and narrow capability to upsurge biliary Cu excretion. It is also known that certain species of sheep have capacity of Copper tolerance and resistance and give a way for cross breeding and selection programs [58]. Cattle has been identified as typical example of ruminant species that could led to tolerant of Cu accumulation and its toxicity in cattle is rare [48]. Actually, Cu deficiency is a fairly common problem in cattle diets globally and higher concentrations of Cu is supplemented up to 35 mg/kg DM which is the maximum level of Cu concentrations determined by European Union, compare to general physiological needs 10 mg/kg DM [39]. In maintainable systems, soil is the chief source of trace minerals for the livestock and farms waste materials can be reprocessed on the paddock [59]. The availability of minerals in various soil types is totally dependent on parent rock concentrations and on soil chemistry; therefore, forages in certain situations may be imbalanced or deficient in certain trace minerals [60–63].

3.7 Conclusion and Recommendations

This chapter has provided detailed information on the benefits of geochemistry in livestock most especially in the adequate maintenance of animal health and animal nutrition. Special highlights were also made on the significant of numerous mineral that have several roles in effective production of livestock. Geochemical evaluation of soil has been identified as a better option for improving livestock productivity in terms of organic nutritional diets while a healthy soil could acts as mother nourishments for crops and indirectly for grazing animals. Also, it was realized that healthy animals deliver greater output when compared to unhealthy animals. Special attention was also laid on geochemistry as the branch of earth science that uses the instrument and principles of chemistry to clarify the mechanism working past present of earth. The various branches like analytical geochemistry, isotope geochemistry, organic and petroleum geochemistry, cosmochemistry (it is the class of technical questions being spoken), environmental geochemistry, and low temperature geochemistry were also highlighted. There is a need to explore the unexplored minerals that are available in different agro-ecological zones of different nations. More experiments need to be carried out regarding the nutrition of agriculture livestock as it deals with the amount of ingested elements by the grazing cattle and sheep.

References

1. Fordyce, F., Selenium Geochemistry and Health. *AMBIO A Journal of the Human Environment*, 36, 1, 94–97, 2007.
2. Niederman, C.N., Blodgett, D., Eversole, G., Schurig, G., Thatcher, C.D., Effect of copper and iron on neutrophil function and humoral immunity of gestating beef cattle. *J. Am. Vet. Med. Assoc.*, 204, 1796–1800, 1994.
3. Stanford University, *Harmful Environmental Effects Of Livestock Production On The Planet' Increasingly Serious,' Says Panel. ScienceDaily*, ScienceDaily, 22 February 2007, <www.sciencedaily.com/releases/2007/02/070220145244.htm>.
4. Meyer, J.A., Casey, N.H., Myburgh, J., The influence of geochemistry on health risks to animals and humans in geographically localised livestock production systems. Short paper and poster abstracts: *38th Congress of the South African Society of Animal Science*, The South African Journal of Animal Science, 2000, is available online at http://www.sasas.co.za/Sajas.html 82.
5. Kwon, M.J., Yun, S., Ham, B., Lee, J., Jheong, W., Impacts of leachates from livestock carcass burial and manure heap sites on groundwater geochemistry and microbial community structure, 2017Published: August 3, 2017, https://doi.org/10.1371/journal.pone.0182579.
6. Infectious Disease News. February 01, 2008. One Health Initiative, One Health Initiative will unite human and veterinary medicine, 2017, Retrieved from http://www.onehealthinitiative.com/index.php.
7. Thornton, T., Geochemistry and the mineral nutrition of agricultural livestock and wildlife. *Appl. Geochem.*, 17, 8, 1017–1028, 2002.
8. Thornton, I., Environmental geochemistry: 40 years research at Imperial College, London, UK. *Appl. Geochem.*, 27, 5, 939–953, 2012.
9. Bradl, H., *Heavy metals in the environment: origin, interaction and remediation*, Elsevier, Academic Press, 2005.

10. Abrahams, P.W., Involuntary soil ingestion and geophagia: A source and sink of mineral nutrients and potentially harmful elements to consumers of earth materials. *Appl. Geochem.*, 27, 5, 954–968, 2012.
11. Reimann, C. and de Caritat, P., Distinguishing between natural and anthropogenic sources for elements in the environment: regional geochemical surveys versus enrichment factors. *Sci. Total Environ.*, 337, 1–3, 91–107, 2005.
12. Kabata-Pendias, A., Dudka, S., Chlopecka, A., Background levels and environmental influences on trace metals in soils of the temperate humid zone of Europe, in: *Biogeochemistry of trace metals*, pp. 73–96, CRC Press, Boca Raton (USA), 2017.
13. Burt, R., Wilson, M., Mays, M., Lee, C., Major and trace elements of selected pedons in the USA. *J. Environ. Qual.*, 32, 6, 2109–2121, 2003.
14. Guillén, M.T., Delgado, J., Albanese, S., Nieto, J.M., Lima, A., De Vivo, B., Environmental geochemical mapping of Huelva municipality soils (SW Spain) as a tool to determine background and baseline values. *J. Geochem. Explor.*, 109, 1–3, 59–69, 2011.
15. Rahman, M., Lee, S.-H., Ji, H.C., Kabir, A.H., Jones, C.S., Lee, K.-W., Importance of mineral nutrition for mitigating aluminum toxicity in plants on acidic soils: current status and opportunities. *Int. J. Mol. Sci.*, 19, 10, 3073, 2018.
16. Khan, Z., II, Ashraf, M., Hussain, A., Evaluation of macro mineral contents of forages: Influence of pasture and seasonal variation. *Asian-Australas. J. Anim. Sci.*, 20, 6, 908–913, 2007.
17. Shimshony, A., Disease prevention and preparedness in cases of animal health emergencies in the Middle East. *Rev. Sci. Tech.-Off. Int. Épizoot.*, 18, 66–73, 1999.
18. Hansen, S. and Spears, J., Impact of copper deficiency in the presence or absence of high dietary manganese on iron status of cattle. *J. Anim. Sci.*, 86, 209–236, 2008.
19. Marai, I.F.M., El-Darawany, A.-H., Ismail, E., Abdel-Hafez, M.A.M., Reproductive and physiological traits of Egyptian Suffolk rams as affected by selenium dietary supplementation and housing heat radiation effects during winter of the sub-tropical environment of Egypt. *Arch. Anim. Breed.*, 52, 4, 402–409, 2009.
20. Humphreys, L.R., *Tropical pasture utilisation*, Cambridge University Press, 2005. Hussain, S.G. and Iqbal, A., *FINAL DRAFT i Bangladesh Agricultural Research Council New Airport Road*, Farmgate, Dhaka-1215 Bangladesh.
22. Baumann, F., S. HE, J., Schmidt, K., Kuehn, P., Scholten, T., Pedogenesis, permafrost, and soil moisture as controlling factors for soil nitrogen and carbon contents across the Tibetan Plateau. *Global Change Biol.*, 15, 12, 3001–3017, 2009.
23. Nael, M., Khademi, H., Jalalian, A., Schulin, R., Kalbasi, M., Sotohian, F., Effect of geo-pedological conditions on the distribution and chemical

speciation of selected trace elements in forest soils of western Alborz, Iran. *Geoderma*, 152, 1–2, 157–170, 2009.
24. Yang, Z., Yu, T., Hou, Q., Xia, X., Feng, H., Huang, C., Wang, L., Lv, Y., Zhang, M., Geochemical evaluation of land quality in China and its applications. *J. Geochem. Explor.*, 139, 122–135, 2014.
25. Rollinson, H.R., *Early Earth systems: a geochemical approach*, John Wiley & Sons, 2009.
26. DiPippo, R., *Geothermal power plants: principles, applications, case studies and environmental impact*, Butterworth-Heinemann, 2012.
27. Plant, J., Smith, D. *et al.*, Environmental geochemistry at the global scale. *Appl. Geochem.*, 16, 11-12, 1291–1308, 2001.
28. Dissanayake, C. and Chandrajith, R., Medical geochemistry of tropical environments. *Earth-Sci. Rev.*, 47, 3-4, 219–258, 1999.
29. Gill, R., *Modern Analytical Geochemistry: an introduction to quantitative chemical analysis techniques for Earth, environmental and materials scientists*, Routledge, 2014.
30. Thode, H., Monster, J. *et al.*, Sulphur isotope geochemistry. *Geochim. Cosmochim. Acta*, 25, 3, 159–174, 1961.
31. Taylor, H.P., The oxygen isotope geochemistry of igneous rocks. *Contrib. Mineral. Petrol.*, 19, 1, 1–71, 1968.
32. Hoefs, J., *Stable isotope geochemistry*, Springer, 2009.
33. Guangzhi, T., *Low-temperature geochemistry*, VSP, 1996.
34. Bordenave, M.L., *Applied petroleum geochemistry*, Technip Paris, 1993.
35. Lewis, G., Geochemistry and animal health, in: *Applied Geochemistry in the 1980s*, pp. 260–269, Graham and Trotman, London, 1986.
36. Cullen, A., Our Pulitzer Prize-winning editorials, 2017. Retrieved from https://sltimesblog.wordpress.com/project-type/winning-work/ Google Scholar.
37. Balmurugan, B., Ramamoorthy, M., Shankar, R., Mineral an Important Nutrient for Efficient Reproductive Health Indairy Cattle. *Int. J. Sci. Environ. Technol.*, 6, 1, 694–701, 2017, 2277–663, 2278-3687 (O).
38. Kumar, S., Pandey, K.A., Razzaque, W., Dwivedi, D.W., Importance of micro minerals in reproductive: Performance of livestock. *Vet. World*, 4, 5, 230–233, 2011.
39. López-Alonso, Trace Minerals and Livestock: Not TooMuch Not Too Little. *International Scholarly Research Network ISRN Vet. Sci.*, vol. 2012, Article ID 704825, 18 pages, 2012.
40. Close, W.H., Trace mineral nutrition of pigs. Meeting production and environmental objectives, in: *EAAP Annual Meeting*Antalya, Turkey, 2006.
41. Suttle, N.F., *Mineral Nutrition of Livestock*, Cabi Publishing, 2010.
42. Babinszky, L., Halas, V., Verstegen, M.W.A., Blanco, J., Kheradmand, H., *Impacts of Climate Change on Animal Production and Quality of Animal Food Products*, 2011-09-09 PY - 2011, N2 - https://doi.org/10.5772/23840 DO - 10.5772/23840 SN - PB - IntechOpen CY - Rijeka Y2 - 2020-03-25 ER –.

43. Gerber, P. and Steinfeld, H., Worldwide growth of animal production and environmental consequences, in: *Trace Elements in Animal Production Systems*, pp. 21–32, 2008.
44. Alexandratos, N., Bruinsma, J., Boedeker, G., Schmidhuber, J., Broca, S., Shetty, P., Ottaviani, M., World agriculture: Towards 2030/2050. Interim report. Prospects for food, nutrition, agriculture and major commodity groups, 2006.
45. Kesler, S.E., Simon, A.C., Simon, A.F., *Mineral resources, economics and the environment*, Cambridge University Press, 2015.
46. Yatoo, M., II, Saxena, A., Deepa, P.M., Habeab, B.P., Devi, S., Jatav, R.S., Dimri, U., Role of trace elements in animals: a review. *Vet. World*, 6, 12, 963, 2013.
47. Sands, D.C., Morris, C.E., Dratz, E.A., Pilgeram, A.L., Elevating optimal human nutrition to a central goal of plant breeding and production of plant-based foods. *Plant Sci.*, 177, 5, 377–389, 2009.
48. Jobling, M., *National Research Council (NRC): Nutrient requirements of fish and shrimp*, Springer, 2012.
49. Alonso, M.L., Montaña, F.P., Miranda, M., Castillo, C., Hernández, J., Benedito, J.L., Interactions between toxic (As, Cd, Hg and Pb) and nutritional essential (Ca, Co, Cr, Cu, Fe, Mn, Mo, Ni, Se, Zn) elements in the tissues of cattle from NW Spain. *Biometals*, 17, 4, 389–397, 2004.
50. Meglia, G., Nutrition and immune response in periparturient dairy cows, Diss. (sammanfattning/summary) Uppsala : Sveriges lantbruksuniv., Acta Universitatis Agriculturae Sueciae. Veterinaria, 170, 1401–6257, 2004.
51. Tame, M., Management of trace elements and vitamins in organic ruminant livestock nutrition in the context of the whole farm system. Institute of Organic Training and Advice (IOTA), Craven Arms, http://www.organicadvice.org.uk/papers/Res_review_4_tame2.pdf, 2008.
52. Soetan, K., Olaiya, C., Oyewole, O., The importance of mineral elements for humans, domestic animals and plants: A review. *Afr. J. Food Sci.*, 4, 80–91, 2009.
53. Delgado, C., Rosegrant, M., Steinfeld, H., Ehui, S., Courbois, C., Livestock to 2020: The next food revolution. *Outlook Agric.*, 30, 1, 27–29, 2001.
54. Alonso, M.L., Benedito, J., Miranda, M., Castillo, C., Hernandez, J., Shore, R., Interactions between toxic and essential trace metals in cattle from a region with low levels of pollution. *Arch. Environ. Contam. Toxicol.*, 42, 2, 165–172, 2002.
55. Blanco-Penedo, I., Cruz, J., López-Alonso, M., Miranda, M., Castillo, C., Hernández, J., Benedito, J., Influence of copper status on the accumulation of toxic and essential metals in cattle. *Environ. Int.*, 32, 7, 901–906, 2006.
56. Mercer, J.F., The molecular basis of copper-transport diseases. *Trends Mol. Med.*, 7, 2, 64–69, 2001.

57. Suttle, N., Lewis, R., Small, J., Effects of breed and family on rate of copper accretion in the liver of purebred Charollais, Suffolk and Texel lambs. *Anim. Sci.*, 75, 2, 295–302, 2002.
58. Kendall, N., Illingworth, D., Telfer, S., Copper responsive infertility in British cattle: the use of a blood caeruloplasmin to copper ratio in determining a requirement for copper supplementation. *BSAP Occas. Publ.*, 26, 2, 429–432, 2001.
59. Hayashida, M., Orden, E.A., Cruz, E.M., Cruz, L.C., Fujihara, T., Effects of concentrate supplementation on blood mineral concentration of growing upgraded Philippine goats. *Anim. Sci. J.*, 75, 2, 139–145, 2004.
60. Coonan, C., Freeston-Smith, C., Allen, J., Wilde, D., *Determination of the major mineral and trace element balance of dairy cows in organic production systems*, vol. 106, pp. 181–184, Publication-European Association for Animal Production, 2001.
61. Owens, K., *Copper, Cobalt and Molybdenum budgets on organic farms in Aberdeenshire* [MS thesis], SAC, 2001.
62. Bowling, P.J., Weller, R.F., Kyriazakis, I., Zervas, G., The yield and quality of plant species grown in mixed organic swards, 2002.
63. Govasmark, E., Steen, A., Bakken, A.K., Strøm, T., Hansen, S., Factors affecting the concentration of Zn, Fe and Mn in herbage from organic farms and in relation to dietary requirements of ruminants. *Acta Agr. Scand., Sect. B-Soil Plant Sci.*, 55, 2, 131–142, 2005.

4

Application in Geochemistry Toward the Achievement of a Sustainable Agricultural Science

Muhammad Akram[1], Charles Oluwaseun Adetunji[2*], S.O. Okonkwo[3], Inamuddin[6], Umme Laila[1], J. Inobeme[4], A. Inobeme[5], Saher Islam[7] and Maliki Munirat[5]

[1]Department of Eastern Medicine, Government College University Faisalabad, Faisalabad, Pakistan
[2]Applied Microbiology, Biotechnology and Nanotechnology Laboratory, Department of Microbiology, Edo University Iyamho, Auchi, Edo State, Nigeria
[3]Department of Chemistry, Osaka University, Osaka, Japan
[4]Department of Geography, Ahmadu Bello University, Zaria, Nigeria
[5]Department of Chemistry, Edo University, Iyamho, Nigeria
[6]Department of Chemistry, King Abdulaziz University, Jeddah, Saudi Arabia
[7]Institute of Biochemistry and Biotechnology, Faculty of Biosciences, University of Veterinary and Animal Sciences, Lahore, Pakistan

Abstract
The world population has been stated to increase tremendously to 9 billion in the year 2050. Therefore, there is a need to explore significant numerous mineral elements that have not be explored before. The earth crust is blessed with numerous essential minerals which could play an essential function in the growth of sustainable agriculture and increase food production as well as the maintenance of a cleaner environment. Therefore, this chapter explores the significance of geochemistry in the identification of useful mineral element that could boost agricultural production and their significant role in the decontamination of heavily polluted soil and water. Special highlights were made on various classifications of geochemistry, chemical composition, as well as their benefits most especially

**Corresponding author*: charliguitar@yahoo.com; adetunjicharles@gmail.com, (ORCID: 0000-0003-3524-6441)

Inamuddin, Mohd Imran Ahamed, Rajender Boddula and Tariq Altalhi (eds.) Geochemistry: Concepts and Applications, (57–72) © 2021 Scrivener Publishing LLC

in the area of sustainable agriculture. Detailed information was provided on the significance of geochemical effects of heavy metals on crops health.

***Keywords*:** Geochemistry, sustainable, agriculture essential minerals, heavy metals, crops health

4.1 Introduction

It has been recognized that there is a great relationship between agriculture and geochemistry. The deeper knowledge of geochemical has been shown as a subject that provides several essential elements that could lead to increase in the development of plant. Some unique geochemical properties involves atmospheric sediment, water, and some essential mineral component that area available in the soil. It has been stated that some essential condition that could lead to great increase of plant includes water, sunshine, water, and temperature as well as the level of soil fertility [1].

Moreover, the level of soil fertility might be linked to the presence of trace metals and some essential constituents available in the earth crust. Also most of these essential elements have been documented to perform a special role in the maintenance of human health. Also, it has been observed that when these elements are present beyond the required level or available in unlimited supply in the soil, it might lead to rapid development of endemic plant diseases. Moreover, for effective function of the soil, there is a need for addition of some agricultural inputs such as pesticides and fertilization. Also, the discharge of waste water and the removal of exhaust gas have a greater influence on the level of soil function most especially in maintenance of the normal soil function [2].

It has been observed that plants have the capability absorb heavy metals and organic chemicals that are available in the soil into the food chain which could lead to significant rise in their content. This might later affect human health. Examining the geochemical features of soil, water, and environment is very important for effective agricultural management of available resources as well as effective planning of these resources [3]. There are some chemical elements that could be withdraw into plant most especially from the soil through diverse means such as polluted water, irrigation water, and agricultural wastes, as well as the polluted sediments which could be achieved through geochemical survey of land quality. It has been stated that the presence of soil elements could affect the growth of plant which could also influence the health of human as well as their lives. Therefore, it is possible to establish the relationship between the development of crops and the quality of the chemical elements available in different products that of agricultural important [4].

It has been observed that geochemical assessments is necessary for evaluating the land quality as well as for assessing the plant growth polluted with heavy metals, excessive residues of pesticides, and organic compounds. The application of geochemical land quality classification is necessary for grading and investigation of environmental, human, and geological factors which all have a relationship to the activity of mankind in terms of land usage [1, 5, 6]. Therefore, this chapter intends to provide the significance of important of geochemistry and agriculture, most especially the application of essential element for the achievement of sustainable agriculture. Special highlights were made on various classifications of geochemistry, chemical composition, as well as their benefits most especially in the area of sustainable agriculture. Detailed information was provided on the significance of geochemical effects of heavy metals on crops health.

4.2 General Overview on the Utilization of Geochemistry and Their Wide Application on Agriculture

Geochemistry is one of the most important and widely classified systems most especially because of their wide application in agriculture. In this study chapter, more emphasis will be laid on essential element mostly obtained from minerals which are rock based, atmosphere, water, and living things. The main role of geochemistry is the origination of chemical element from the nature. Geochemistry is a vast field is divided naturally into different levels. They are classified into different types like analytical geo-chemistry, photo geo-chemistry, isotope geo-chemistry, regional geo-chemistry, cosmo-chemistry, igneous geo-chemistry, petroleum geo-chemistry, aqueous geo-chemistry, and organic geo-chemistry. The application of geochemistry have also been established to perform an essential function in the monitoring, management of environmental factors, earth exploration of petroleum and metal from the earth crust. The role of geo-chemistry could also helpful in understanding the most common environment issues like water pollution, depletion of ozone, and global warming.

4.2.1 Classification

Geochemistry divided into different types that include:

Reginal geo-chemistry:
This involves researches on hydrological, mineral, and environmental study [7].

Isotope geo-chemistry:
This involves researches on element concentration and isotope which present on the surface of earth [8].

Organic geo-chemistry:
This involves researches on those compounds which are derived from living organism [9]. Some other examples include like photo geochemistry, cosmochemistry, biogeochemistry, and aqueous geochemistry.

4.2.2 Chemical Composition of Rocks

The chemical composition of rocks mostly contains fluorides, sulfides, chlorides, and oxides and their amount which is present in the rocks are usually less than 1%. From the research of different scientist, it has been discovered that almost 47% mostly oxygen is present in the earth crust. Mostly the oxygen is available in the form of oxides which includes oxides of iron, silica, alumina, carbonate of calcium, sodium, magnesium, and potassium. Also, these oxides could also exist in the form of sodium carbonate and potash in the form of potassium carbonate which could combine form feldspars to form leucite, muscovite, and nepheline, respectively. The formation of apatite occurs when lime combine with phosphoric acid. Different components available in the soil show different characteristic such as biological, chemical, and physical. Almost 12 different elements like nickel, iron, calcium, silicon, aluminium, sodium, potassium, chromium, manganese, and magnesium were all available. The soil reginal geo-chemistry shows that minerals mostly the major elements are present in the soil [10].

4.2.3 Effect of Some Beneficial Minerals in Agriculture

Inorganic minerals are utilized by plants for their growth and development. Complex interactions have been recognized to occur between rock mineral weathering, decaying organic matter, organisms, and microbes with the soil. The plant possesses the capability to store these inorganic minerals in the roots of numerous plants as ions. Numerous factors have been identified to influence plant nutrient uptake. Ions can be easily available to roots, or the other elements present as well as the soil could "tie" them together. Soil that has a high pH (alkaline) or too weak (acid) renders the plants inaccessible to minerals [11]. The term fertility could be referred to the potential of plants with essential nutritive contents that will enhance the adequate development of plant in adequate proportion. Nutrition could be define as the interrelated phases that area involved in the absorption of nutrition or essential foods are assimilated mainly by the living microorganism for adequate development and replacement of human tissue. There is a need to increase the level of soil fertility through the provision of adequate nutrients in adequate proportion through the provision of adequate amount of mineral nutrients.

Numerous attempts have been devoted to the development of fertilizer that could be utilized for adequate amelioration of high level nutrient shortage available in the soil. The application of hydroponics have been identified as a system that involve the supply of essential nutrient and mineral to plant for adequate development so as to compensate the loss of nutrient in the soil. This system encourages adequate supply of essential nutrients to the plant. It has been recognized that plants needs essential elements which might entails macronutrients, micronutrients, and trace elements which all have numerous essential function they play toward the supply of essential nutrient to the plant [12]. The essential elements have been recognized to be very important most for all plants. A typical example involves the application of cobalt has been shown as a crucial element important for the fixing of nitrogen available in the legumes. The deposit of the silicon in the cell wall has been validated to improve tolerance of plant to drought and heat as well as improve the level of plant to the adverse effect of infection from bacteria, insects, and fungal. Silicon has been recognized as an essential constituent that could minimize the amount of poisonous level some essential elements [13].

A more balanced plant nourishment approach might not be inadequate to nourishment that are necessary for survival but would include mineral elements that are supplied at the rates that are beneficial for optimum growth of plants. It has been discovered that some essential elements possess the capability to remove pollutants with the help of some recent advances in analytical chemistry [14]. In agriculture, the application of industrial humates has been documented. Also, different variations have been observed between soil humic acids and marketable humates. This investigation was therefore carried out a study in order to assess the geochemical properties and effect of selected profitable humates, fulvic acid, and extracted humic acid obtained from non-lignite and lignite deposits on the plant production. The application of Nuclear Magnetic Resonance spectrometry (^{13}C-NMR) and infrared (IR) spectroscopic technique were used in evaluating their spectral characteristics.

Also, the maize plants that were grown in the occurrence of marketable humates in the greenhouse showed that the humates have an impact on the growth of plant [12]. Commercial humates composed mainly of HA, together with small quantities of FA. Humic acid portions from saleable humates were comparable in carbon and hydrogen constituents but lignite humate, humic acids were higher in nitrogen content when compared to non-lignite humate isolated HA. Lignite-HA also displayed higher carbohydrate content in comparison to the non-lignite HA. By nature, the latter had become more aromatic. The IR spectroscopy confirmed the existence of non-lignite of greater aromatic content in humic acids when compared to lignite deposits. Moreover, the result obtained from the commercial humates could also be comparable to the HA present in the soil and their impact on corn plant production was also established [15].

4.2.4 Beneficial Mineral Nutrients That are Crucial to the Development of Plants

These are the lists of essential mineral nutrients that are essential to the development of plants The absence of these mineral elements may affect plants which may make plant to exhibit growth anomalies, deficiency symptoms and slower development. Nitrogen is a vital part of plant life-saving proteins, hormones, chlorophyll, vitamins, and enzymes. The nitrogen metabolism is an important condition in the growth of stems and leaves (vegetative growth). So much will delay the fruiting and flowering. Deficiencies can lower yields, cause leaf yellowing, and poor growth. Phosphorus is required for seed germination, protein development, photosynthesis, and nearly different forms of plant metabolism and development. This is important for the creation of flowers and fruits. Low pH (<4) results in a chemical lock up of phosphate in organic soils. Symptoms of deficiency are purple stems and leaves, and delays in maturity and development. Fruit yields and flowers are small. Premature fruit and flowers drop regularly. Phosphorus must be added near the roots of the plant for use by the plant. Large phosphorous applications without sufficient zinc levels can cause a deficiency in zinc [10].

Potassium is required to form carbohydrates, sugars, carbohydrates, synthesis of protein, and division of cells into the various sections of the plant. It play functional role in the improvement of the level of the oil constituent, fruit development, and prevention of several deficiency in plant. Magnesium is required to generate carbohydrates, sugars, and fats for the functioning of plant enzymes [16]. Calcium stimulates enzymes and is a basic component of cell walls; it influences the transport of water in cells and is useful for the growth and division of cells. Calcium is usually immobile once it is deposited in plant tissue (cannot be translocated) so the supply for growth must be constant. Deficiency causes the branches, bulbs, and roots to stall new growth [17].

4.2.4.1 Micronutrients

Iron plays a crucial role toward chlorophyll development in plant. It is important for plant growth most especially the young ones. Deficiencies of young leaves are light leaf color in affected plants. Moreover, whenever iron is leached brings about it loss and stored in the lower parts of the soil. The presence of iron element is made unavailable to plants under high pH (alkaline) conditions. The iron can be plentiful but unavailable when soils are alkaline. Applications of an iron chelate-containing acid nutrient solution, kept in soluble form, will correct the issue [16]. Manganese plays an active role during the process of photosynthesis, respiration, and metabolism of nitrogen enzymes, respectively. Deficiency in young leaves may exhibit a green vein network on a light green environment which is an indication of deficiency due to iron element. The light green sections get white in the advanced stages, and leaves are shed while the veins can appear brownish, black, or grayish spots. Plants exhibit deficiency symptoms often in acidic or alkaline soils. Manganese can be present in strongly acid soils, which might lead to high level of contamination.

Boron has been identified to play a crucial role in the cell wall formation, for the integrity of membranes, for the absorption of calcium and can help to translocate sugars. Boron affects a total of 16 plant functions. Such functions include flowering, cell division, and germination of seeds, fruiting, water connections, and hormone movement. Boron must be present all along the plant's life. The affected plant section might become colored with brown spots, broken and flecked. Zinc is a part of a functional cofactor or an enzyme. Metabolism, protein synthesis, and elongation of internodes (stem growth) are important for carbohydrates. Deficient plants have irregularly chlorotic areas with mottled leaves [18]. Copper is found in plant roots and plays a vital role in the metabolism of nitrogen. It is a component of many enzymes and is part of carbohydrates and protein-using enzyme systems. It deficiencies could lead to development of brown spots on the terminal leaves. Molybdenum is a functional enzyme component, which converts nitrates to ammonia. It has been observed that the absence protein synthesis could block the development of plants. It is also needed by root nodule (nitrogen fixation) bacteria. Seeds cannot fully develop, if the plants lack the presence of molybdenum which can also led to a nitrogen deficiency. The presence of pale green color with cupped margins or rolled are indicators of deficiencies on the leaves.

Chlorine plays a role in osmosis (water movement or cell solutes) and plays a vital role in the process of photosynthesis. Symptoms of chlorine entail chlorosis of the leaves while odors can be decreased in numerous component of plants section. The presence of chloride which could exist in the form of chlorine ionic compound could be utilized by plants when present in soluble forms and could be leached to the lower regions of the soil. Moreover, if the rates are too high, some plants may display signs of toxicity [19]. Nickel is vital for the breakdown of urea by the enzyme known as urease so as to release nitrogen gas into a valuable form in the plant. Nickel is required to absorb the iron. Cobalt has been documented as a significant element needed in legumes and in non-legumin root nodules for nitrogen fixation. For nitrogen fixation the quest for cobalt is much greater than for nutrition for ammonium. Deficient levels may lead to symptoms of a nitrogen deficiency.

Soil chemistry primarily focused on soil chemical reactions that could lead to paedogenesis or that could affect plant growth. Since then, concerns have been increasing about contamination of organic and inorganic soil from environmental emissions and possible threats to ecological health and environmental safety. Consequently, the focus on soil chemistry has changed agricultural and pedology soil science to an emphasis on soil science for the climate. Knowledge of environmental soil chemistry constitutes a major importance for predicting the fate of pollutants and the mechanisms through which they are initially released into the soil. Once a product is introduced to the soil layer, there may be various chemical reactions that may increase or decrease toxicity to the pollutants. Such reactions include adsorption/desorption, polymerization of precipitation, dissolution, complexation, and reduction of oxidation [16]. The proper understanding of these processes could help the Engineers and scientists that are involved in environmental remediation in accurate forecasting of the fate and toxicity of pollutants and provide the information necessary in the establishment of more scientifically accurate and cost-effective remediation strategy [19, 20].

4.3 Role of Geochemistry in Agriculture

The role of geo-chemistry is effective in the field of agriculture, they also play important role in human health. As we discussed already, the composition of soil is rich of different component which improve the efficacy of soil. Some components are good for the soil but some show toxic effect. For the growth and development of plant, atmosphere, water, and soil gave nutrients which are crucial for the development of plants. The color of the plant and their growth not only depend on air, water, temperature, but the main role which plays in the growth of plant is the fertility of soil. In fertilizer, soil contains many earth elements which are rare, and some are trace. In trace elements that include potassium, zinc, iron, calcium, magnesium, chloride, and many more, these elements also useful not only for animal but also for the human health. The occurrence of disease depends mostly on the presence of element either these elements are in less amount or excess. From the previous year study, it was concluded that soil is mostly affected by the pesticides spray, waste water-discharge, also affected from industry by gas and slag exhaust [5].

The presence of heavy metal and biochemical increases in soil by absorption of plant through the chain of food, and this method useful for the re-structuring agriculture, for the management of soil, helpful for the quality improvement of soil. These chemical elements which are essential for the growth of plant not only taken from soil but it also come from atmospheric dust and water irrigation. From the study analysis, it was concluded that the soil which is useful for agriculture purpose and the growth of plant is mostly occurs in those soil which are riches source of nutrients. From the analysis, they selected the soil which is useful for growth or not for growth purpose [1].

4.4 Geochemical Effects of Heavy Metals on Crops Health

McDowell *et al.* briefly reviewed that geographical trace elements diseases distribution in animals determined by soil-plant relationship in tropical regimes. Generally, small ratio of trace minerals in soils is taken up by plants. Soil, plant species, maturity level, yield, pasture handling, and climate all affect forage trace minerals concentrations. Heavy metals release to environment is noticeable increasing environmental problems affecting the balance at ecosystem level. The plants early life cycle phases such as seedling and seed germination are broadly pretentious by heavy metals, disturb individual plant life, population, and communities [10].

There are different ways of causing heavy metals toxicities, counting in storage food mobilization, plant osmoregulation and photosynthesis depends on soil pH, plant species, and type of heavy metals and its kind and mobility of heavy metal. Plants use various mechanisms like antioxidant system activation, production of proline, acetyl salicylic acid, and heat shock proteins, and others for overcoming the metal induced toxicities. Heavy metals contaminated soils can cause high level of toxicity in both plants and animals and considered as disparaging ingredients for the process of photosynthesis. The foremost factors responsible for the accessibility of heavy metals in soils are clay content, organic matter, and soil pH. The mature compost that is humified organic matter also responsible for decreasing the heavy metals bioavailability although increase the residual heavy matter portion in soils. Furthermore, the extensive superficial region of clay minerals causes greater prevention of movement heavy matter in soil. It is investigated that as the pH of soil increases, the heavy metals' immobilization also increases [4].

Biochar presentation to soil causes in increase pH in the soil. This could be a consequence or presence of pH that are available in heavy metal immobilizations. Subsequently due to non-degradable nature of heavy metals can root long period harmful effects on ecosystem. It has been demonstrated that some utmost toxic metalloids exist in soil. Cadmium is known as phytotoxicant and is non-essential for metabolic activities. Cadmium possesses the carcinogenic, cytotoxic, and mutagenic properties due to its long half-life. Coal burnings, industrial effluents, ore smelting, and arsenic containing insecticides and herbicides expel a toxic metalloid as to environment. Mercury is a strong toxic heavy metal mainly release from mining actions. The presence of high amounts of heavy elements in soil cause harmful effects on plants like chlorophyll destruction, growth retardation, biochemical processes disorders, reproductive disorders, and mutations. Heavy metals inhibit the seed germination and seed development. Heavy metals negatively affect the seed germination and seedling development (Table 4.1). Hg is well known about influencing the seed growth. The chlorophyll central atom that is Mg is replaced by heavy metals like Cd, Cu, Hg, Ni, and Pb decline and prevents photosynthesis. Cadmium is known as inhibitor of germination by many studies. Likewise, Cd stressed plants has been showed lesser root and coleoptile growth. Cd inhibits the carbohydrates hydrolysis and translocation of hydrolyzed sugars causing seedling growth reduction. Another toxic metalloid is arsenic, famous for

Table 4.1 Effects of heavy metals on different crops (forages).

Species	Heavy metals	Effects	References
Albazia lebbeck	Hg	A noteworthy steady lessening in seed germination observed yonder 1 mM	[21]
Arachis hypogea	Cd, Pb, Cu	Reduction in seed germination	[22]
Brassica juncea	Cd	Slight effect on seed germination	[23]
Cajanus cajus	Hg	Slight effect on seed germination	[24]
Daucus carota	As	Lessening in seed germination was detected at almost 10 mg L^{-1}	[25]
Eruca sativa	Cu	Lesser effect on seed germination	[26]
Festuca rubra	Cd	No seed germination reduction observed at 25 but decrease in germination was detected 50 mg L^{-1}	[27]
Glycine max	Cd	As concentration of Cd increases seed germination decreased over 5 m/kg while as concentration decrease the deleterious effects decreases at 1 mg/kg	[28]
Helianthus annus	Cd	Decrease in germination at 2–10 mg L^{-1}	[29]
Lens culinaris	Pb	A noteworthy drop in germination was detected from 0.5 mM	[30]

(*Continued*)

Table 4.1 Effects of heavy metals on different Crops (forages). (*Continued*)

Species	Heavy metals	Effects	References
Medicago sativa	Cu	Prevent seed germination by 40% with almost 40 mg L^{-1}	[31]
Oryza sativa	Cd,	Cd displays a noteworthy drop in rice seed germination index	[32]
	Cu	The seeds germination reduced as the concentration improved (0.2–1.5 mM)	[33]
Triticum aestivum	Cd	Cd displayed poisonous attribute at 5 mg L^{-1} in shoots and root growth	[34]
	Cu	The seeds germination decreased from 5 mg L^{-1}	[35]
Phaseolus vulgaris	Co, Cr	Decrease growth of seed at high dose containing 10 to 2 M	[36]
Zea mays	Al	Decerase in germination of seedat 50 mg L^{-1}	[37]
	Cu	Augmented seed germination at 0.1 mM	[33]

suppressing plant growth, that exhibit noteworthy negative effects on shoot and root lengths and their biomass of wheat seedlings [3].

Lead is also a toxic heavy metals excreted from many industrial activities. Pb-treated pea seedlings showed 50% reduction of amylase ensuring high starch contents. The increase in production of hexokinase (HXK) and decrease in 6-phosphogluconate dehydrogenase (G6PDH) assembly in the shoot and the root section of plant which validates those hexoses could be ingested during the process of glycolysis when compared to the pathway of phosphate pathway. Correspondingly, Pb prevents the movement of starch by reducing the hydrolyzing enzymes especially starch available in *Brassica campestris*. Cr has been documented to prevent inhibits the process involved in the germination of seeds through harmful influence on degrading enzymes. The *Zea mays* were salted with different soil levels of copper, nickel, lead, and zinc, resulting in decrease in photosynthesis due to these heavy metals and decline in growth for the duration of prolonged contact periods [20].

4.5 Conclusion and Recommendations

Therefore, this chapter explores the significance of geochemistry in the identification of useful mineral element that could boost agricultural production and their significant role in the decontamination of heavily polluted soil and water. The authors also suggested that many trace and major elements are present in the rocks and these elements important for the fertilization of soil while careful attention needs to be placed on some them that are toxic in nature. There is a need for multi-interdisciplinary research between different researchers across various discipline such as chemistry, geology, agriculture, microbiology and many other disciplines to identify more essential minerals that could boost increase in agricultural production and intensification in food production such as the application of some unutilized mineral for massive production of fertilizer and their application as insecticides and pesticides, thereby reducing the incidence of agricultural pest and diseases.

References

1. Ma, L. and Li, Z., Review of application of geochemistry in agriculture. *SEG 2019 Workshop: Geophysics for Smart City Development*, 29-31 July 2019, Society of Exploration Geophysicists, Beijing, China, 2019.
2. Aktar, M.D., Sengupta, D., Chowdhury, A., Impact of pesticides use in agriculture: their benefits and hazards. *Interdiscip. Toxicol.*, 2, 1, 1–12, 2009.
3. Chibuike, G.U. and Obiora, S.C., Heavy Metal Polluted Soils: Effect on Plants and Bioremediation Methods. *Appl. Environ. Soil Sci.*, 2014, Article ID 752708, 12, 2014. https://doi.org/10.1155/2014/752708.
4. Wuana, R.A. and Okieimen, F.E., Heavy Metals in Contaminated Soils: A Review of Sources, Chemistry, Risks and Best Available Strategies for Remediation. *Int. Sch. Res. Notices*, 2011, Article ID 402647, 20p, 2011. https://doi.org/10.5402/2011/402647.

5. Gentile, A.R., Soil erosion - European Commission - Europa EU ec.europa.eu › environment › archives › soil › pdf ›. 2, 2000, https://ec.europa.eu/environment/archives/soil/pdf/vol5.pdf.
6. Pereira, V. and Martinho, D., Exploring the Topics of Soil Pollution and Agricultural Economics: Highlighting Good Practices. *Agriculture*, 10, 1, 24, 2020, https://doi.org/10.3390/agriculture10010024.
7. Garrett, R., Reimann, C. *et al.*, From geochemical prospecting to international geochemical mapping: a historical overview. *Geochem.: Explor. Environ. Anal.*, 8, 3-4, 205–217, 2008.
8. Kendall, C. and Caldwell, E.A., Fundamentals of isotope geochemistry, in: *Isotope tracers in catchment hydrology*, pp. 51–86, Elsevier, London, United Kingdom, 1998, https://doi.org/10.1016/B978-0-444-81546-0.50009-4.
9. Killops, S.D. and Killops, V.J., *Introduction to organic geochemistry*, John Wiley & Sons, Blackwell Publishing Ltd, New Jersey, USA, 2013.
10. El-Anwar, A., Mekky, E.A., Abdel, H.S., Wahab, W., Geochemical characteristics of agricultural soils, Assiut governorate, Egypt. *Bull. Natl. Res. Cent.*, 43, 41, 2019, https://doi.org/10.1186/s42269-019-0080-3.
11. Padhan, D., Kundu, R., Sen, A., Adhikary, S., *Current Research in Soil Fertility*, AkiNik Publications, Delhi, India, 2019.
12. Yakimenko, O. and Terekhova, V., Humic Preparations and the Assessment of Their Biological Activity for Certification Purposes. *Eurasian Soil Sci.*, 44, 11, 1222–1230, 2011.
13. Kaur, S., Kaur, N., Siddique, K., Nayyar, H., Beneficial elements for agricultural crops and their functional relevance in defence against stresses. *J. Arch. Agron. Soil Sci.*, 62, 7, 2016.
14. Singh, A.L. and Mann, A., Water and salinity stresses and iron and Zinc biofortification in Groundnut: Physiological and molecular approaches for development of climatic resilient crops. *Proc National Seminar of Plant Physiology*, 12-14 Dec 2012, Recent Advances in Plant Nutrition, 2012.
15. Francioso, O., Montecchio, D., Giocchini, P., Ciavatta, C., Thermal analysis (TG–DTA) and isotopic characterization (13C–15N) of humic acids from different origins. *Appl. Geochem.*, 20, 3, 537–544, 2005.
16. Soetan, K.O., Olaiya, C.O., Oyewole, O.E., The importance of mineral elements for humans, domestic animals and plants. *Afr. J. Food Sci.*, 4, 5, 200–222, 2010.
17. Carroll, J., What Causes Black Spots on Watermelons. *J. Appl. Sci. Res.*, 9, 7, 4765–4780, 2017.
18. Walia, U.S. and Walia, S.S., *Crop Management*, p. 36, Google Books Result, Scientific Publisher, India, 2015.
19. Tchounwou, P., Yedjou, C., Patlolla, A.K., Sutton, D.J., Heavy Metals Toxicity and the Environment. *Technology and Engineering, EXS*. 101, 133–164, 2012.

20. Mohamed, W.S., Ismail, E.A., Zaki, R., Kamel, A., Regional assessment of heavy metals pollution in irrigated soils, west El Minia district, Egypt. *Egypt. J. Geol.*, 59, 101–112, 2015.
21. Iqbal, M.Z., Shafiq, M., Athar, M.J.A.i.e.r., Phytotoxic effects of mercury on seed germination and seedling growth of Albizia lebbeck (L.) Benth. (Leguminosae). *Adv. Environ. Res.*, 3, 3, 207–216, 2014.
22. Abraham, K., Sridevi, R., Suresh, B., Damodharam, T.J.A.J.o.p.s., and research. Effect of heavy metals (Cd, Pb, Cu) on seed germination of Arachis hypogeae L. *Asian J. Plant Sci. Res.*, 3, 1, 10–12, 2013.
23. Srivastava, S., Srivastava, A.K., Suprasanna, P., D'souza, S.J.J.o.e.b., Identification and profiling of arsenic stress-induced microRNAs in Brassica juncea. *J. Exp. Bot.*, 64, 1, 303–315, 2013.
24. Patnaik, A. and Mohanty, B.J.A.o.B.R., Toxic effect of mercury and cadmium on germination and seedling growth of Cajanus cajan L. (pigeon pea). *Ann. Bio. Res.*, 4, 3, 123–126, 2013.
25. Dutta, P., Islam, M., Mondal, S.J.J.o.P. P., and Pathology. Interactive effect of arsenic stress and seed phytate content on germination and seedling development of different vegetable crops. *J. Plant Physiol. & Pathol.*, 2, 2, 2, 2014.
26. Zhi, Y., Deng, Z., Luo, M., Ding, W., Hu, Y., Deng, J., Li, Y., Zhao, Y., Zhang, X., Wu, W.J.A.J.o.P.S., Influence of Heavy Metals on Seed Germination and Early Seedling Growth in Eruca sativa Mill. *Am. J. Plant Sci.*, 6, 05, 582, 2015.
27. Vázquez de Aldana, B., Gundel, P.E., Garcia Criado, B., García Ciudad, A., García Sánchez, A., Zabalgogeazcoa, I.J.G., f. science, Germination response of endophytic Festuca rubra seeds in the presence of arsenic. *Grass and Forage Sci.*, 69, 3, 462–469, 2014.
28. Wan, M., Wang, M., Zhou, F., Yang, L.J.J.o.H.U., Effects of arsenic on seed germination of mung bean and black soybean. *J. Environ. Sci.*, 3, 006, 2013.
29. Imran, M.A., Khan, R., Ali, Z., Mahmood, T.J., II, Toxicity of arsenic (As) on seed germination of sunflower (Helianthus annuus L.). *Int. J. Phys. Sci.*, 8, 840–847, 2013.
30. Cokkizgin, A. and Cokkizgin, H., Effects of Lead (PbCl2) Stress on Germination of Lentil (Lens culinaris Medic.) Cultivars. *Afr. J. Biotechnol.*, 5, 2, 205–211, 2010.
31. Aydinalp, C. and Marinova, S.J.B.J.o.A.S., The effects of heavy metals on seed germination and plant growth on alfalfa plant (Medicago sativa). *Bul. J. Agric. Sci.*, 15, 4, 347–350, 2009.
32. He, J., Ren, Y., Chen, X., Chen and e. safety, H.J.E., Protective roles of nitric oxide on seed germination and seedling growth of rice (Oryza sativa L.) under cadmium stress. *Ecotoxicol. Environ. Saf.*, 108, 114–119, 2014.
33. Ahsan, N., Lee, D.-G., Lee, S.-H., Kang, K.Y., Lee, J.J., Kim, P.J., Yoon, H.-S., Kim, J.-S., Lee, B.-H. J. C., Excess copper induced physiological and proteomic changes in germinating rice seeds. *Chemosphere*, 67, 6, 1182–1193, 2007.

34. Ahmad, I., Akhtar, M.J., Zahir, Z.A., Jamil, A.J.P.J.B., Effect of cadmium on seed germination and seedling growth of four wheat (Triticum aestivum L.) cultivars. *Pak. J. Bot.*, 44, 5, 1569–1574, 2012.
35. Singh, D., Nath, K., Sharma, Y.K.J.J.o.E.B., Response of wheat seed germination and seedling growth under copper stress. *J. Environ. Biol.,* 28, 2, 409, 2007.
36. Zeid, I.J.B.P., Responses of Phaseolus vulgaris chromium and cobalt treatments. *Biol. Plant.*, 44, 1, 111–115, 2001.
37. Nasr, N.J.M.R.J.o.E.S. and Toxicology, Germination and seedling growth of maize (Zea mays L.) seeds in toxicity of aluminum and nickel. *MRJEST*, 1, 5, 110–113, 2013.

5

Geochemistry, Extent of Pollution, and Ecological Impact of Heavy Metal Pollutants in Soil

Abhiroop Chowdhury[1,2*], Aliya Naz[3] and Diksha Sharma[2]

[1]Jindal School of Environment and Sustainability, O.P. Jindal Global University, Sonipat, Haryana, India
[2]Department of Chemistry, Manipal University Jaipur, Dehmi Kalan, Jaipur-Ajmer Express Way, Jaipur, Rajasthan, India
[3]Department of Environmental Science and Engineering, Indian Institute of Technology, (Indian School of Mines) Dhanbad, Jharkhand, India

Abstract

Industrialization has been propelling human race to the epitome of success. But, pollution is also a looming truth that is a major concern to human health, ecosystem, and overall existence of humans at large. Heavy metal contamination is a subject of concern for the geochemists and environmentalists alike, as they are difficult to degrade and have potential to be biomagnified in the food chain, shows toxicity in cellular metabolism, and eventually impacts human health. Water pollution has been the major thrust area of research for environmental geochemists until the recent reported cases of soil contamination across the globe. But, unlike water, soil is a heterogenous medium with no reported permissible levels of metals and hence understanding the impact of pollution in soil has always baffled academicians and policy-makers alike. In this chapter, different cases of heavy metal contamination in edaphic phase have been addressed using Ecological Risk Index–based empirical modeling. The geochemistry of Arsenic, Cadmium, Chromium, Copper, Lead, Nickel, and Zinc has been discussed with special emphasis on its impact on biosphere. It is concluded that mining activity and industrial waste disposals have the major role in soil contamination with "considerably high" or "very high" potential impact on ecosystem.

**Corresponding author*: abhiroop.chowdhury@gmail.com; achowdhury@jgu.edu.in; orcid.org/0000-0001-6985-0722

Inamuddin, Mohd Imran Ahamed, Rajender Boddula and Tariq Altalhi (eds.) Geochemistry: Concepts and Applications, (73–92) © 2021 Scrivener Publishing LLC

Keywords: Bioaccumulation, biomagnification, ecotoxicology, geochemistry, heavy metals, ecosystem, soil, pollution

5.1 Introduction

Nineteenth and twentieth century is the "era of industrialization" in the history of human civilization that has witnessed exponential technological advancement from steam engines to the current age of cybernetics. But, this industrialization and development are adversely affecting nature by contaminating, polluting the pristine environment, and leaving a negative environmental footprint. Mining being the core of human development, it is impractical to assume cessation of pollution overnight following spartan conservation policies wherein all industrial activities are terminated to ensure a safe environment. In the context of developing countries like India, it is also not plausible to restrict mining, as raw materials are mandatory to drive the wheel of industrial development. Hence, as argued by Sarwar *et al.* (2010, 2017) [1, 2], it is challenging to avoid contaminants in the environment-ecosystem nexus due to mining and industrial activities. Contrary to organic pollutants, metals in the environment being non-biodegradable, there is a need for technological interventions to restrict deleterious heavy metals from reaching the food chain or hampering natural ecological balance. Heavy metals tend to disrupt the enzymatic pathways at the cellular level and get bioaccumulated in different trophic levels [3–6]. Because of these reasons, mining activities have received serious attention regarding their potential to destabilize the environment-human nexus [7–11].

Metal pollution in soil can have multitude of impact on not only human but also on the ecosystem as large [12, 13]. The metal pollutants can be bioaccumulated in the tissue system of producers (macro and microphytes) and can be passed through the food chain to the higher trophic levels resulting in problems of biomagnification and can impart toxicity by impacting the cellular metabolic pathways [14, 15].

Understanding the source apportionment of heavy metal contaminants in soil is important for environmentalist and policy makers, prior to the formulation of any policy level interventions. Several case studies have delved into the metal contamination status in the edaphic phase throughout the earth and this chapter deals with a systematic review of the metal pollution status in soil/sediment across the globe due to various activities like mining, agriculture, industrial pollutant output, or other anthropogenic interventions. Here, an index-based approach is used to model the pollutant concentration to reveal their impact on the ecosystem and basic concepts on ecotoxicology, contaminant transfer in the food chain has also been elucidated.

5.2 Material and Methods

5.2.1 Review Process

Literatures in the major scholarly search engines like Google Scholar and ScienceDirect have been accessed to download reverent articles on metal pollution profiles in soil across the globe that are impacted by various anthropogenic pressures like industrialization, agricultural practices, and other causes. The available data on metal pollution profiles from available case studies has been further modeled to understand their impact on the ecosystem.

5.2.2 Ecological Risk Index

Ecological Risk Index (ERI) is an index-based empirical modeling to evaluate the extent of pollution and its impact on the ecosystem. In this method, the concentration of contaminants in the soil is compared with the background values of the similar edaphic phases prior to the industrialization era. Then, the variations in the toxicity of the contaminants are taken into account by equating an already established toxic response factor through the empirical model.

ERI is the summation of contamination factor after computing the toxic response factor for each metal.

$$\text{Contamination factor (CF)} = [C]_{sediment}/[C]_{shale.........} \quad [4–6, 13]$$

where $[C]_{sediment}$ = metal concentration in sediment; $[C]_{shale}$ = average shale value of earth crust. The pollution levels evaluated through C_f values are

CF = 0: none; CF = 1: none to medium; CF = 2: moderate; CF = 3: moderate to strong; CF = 4: strongly polluted; CF = 5: strong to very strong; CF = 6: very strong [4–6, 13].

The shale values (mg/kg) are Lead = 20, Manganese = 850, Nickel = 68, Cadmium = 0.3, Chromium = 90, and Zinc = 95, as per the reported concentrations [16, 17].

Ecological risk factor (ER) can be calculated as

$$ER = TI \times CF^{i}$$

Here, *TI* is Toxic Response Factor for i^{th} element while CF is the contamination factor of ith element. *TI* for individual metals are Lead = 5, Cadmium = 30, Chromium = 2, Zinc = 1, Nickel= 5, and Manganese = 1, as reported [18].

The Ecological Risk Factor values identify the level of pollution due to the impact of heavy metals. So, values are <40 = Low ecological risk, 40–80 = moderate potential ecological risk, 80–160 = considerable potential ecological risk, 160–320 = high potential ecological risk, and ER > 320 = very high ecological risk [4–6, 13].

ERI is summation of all the ER values of individual toxic metals and can be equated as

$$ERI = \sum_{i=1}^{n} ER$$

The value of ERI indicates the overall pollution level of the investigated area. Hence values, <150 indicates "low ecological risk", values between 150 and 300 indicates "moderate ecological risk", values between 300 and 600 indicates "considerable ecological risk", and ERI values greater than 600 indicates "very high ecological risk" [4–6, 13].

5.3 Toxic Heavy Metal and Their Impact to the Ecosystems

5.3.1 Arsenic

The symbol of Arsenic is "As". The atomic number of As is 33. The atomic mass of Arsenic is 74.92. It has an atomic density of 5.727 g/cm^3. It is found in 15 group, also known as Nitrogen group of the periodic table. Naturally, Arsenic originates from geothermal fluids and volcanic emissions. Arsenic mostly exist in +3 or +5 oxidation states [19].

This metalloid generally comes in the environment due to weathering of parental rocks. Some soils have high concentration of Arsenic like shales, clays, and phosphorites [20]. The mean concentration of Arsenic is low in air but increases in urban areas and mining districts due to anthropogenic interventions [21].

The most toxic arsenic species are arsenite and arsenate. Due to the high persistence of arsenic in the environment, it is responsible for environmental as well as health problem. The common chemical species of Arsenic are As (+3) and As (+5), of which As^{+3} is more toxic. It is responsible for several health problems in human like thickening of dermal layer, discoloration of the skin, vomiting, nausea, stomach pain, partial paralysis, blindness and can cause Arsenicosis [19, 22].

The As mobility in soil is directly proportional to its concentration but it is inversely correlated to the time of exposure. Plants bioaccumulate "As" passively with the water flow. There are macrophytes which have high tolerance to the "As" concentration in their tissue; hence, they do not show any physiological or anatomical abnormalities in presence of the metalloid, and they are generally termed as hyperaccumulators [22].

5.3.2 Cadmium

It is discovered in year 1817 and is identified by symbol of "Cd". Atomic number and mass of this transition metal is 48 and 112.41u, respectively, with atomic density of 8.65 g/cm^3.

Cadmium comes into the environment-ecosystem nexus through both geogenic and human linked pollution sources. Cadmium is highly toxic while it does not play any important role in biological functions. The average Cd-concentration in rocks ranges between 100 and 11,000 μg/kg [4, 19].

The most toxic form of cadmium is +2. It is known for its toxicity in biota. Though this element is not required by the cellular metabolic processes of plants, it is observed that it can be readily translocated to the higher plant tissues like leaves, fruits, and even seeds [22]. Cd also has a high propensity to be bioaccumulated in benthic organism. The main mode of toxicity shown by Cadmium is due to its potential to replace Zinc from key catalytic sites of a particular class of enzymes like carbonic anhydrases as investigated in some marine diatoms. Zn^{+2} is replaced by Cd^{+2} because both are present in a same group having similar d10 configuration. Their ligands and coordination numbers are also similar though the ionic radius of Cd^{+2} is slightly greater then Zn^{+2} [23]. That's the main reason for the high bioaccumulation of Cd in biological system and found to be biomagnified in higher trophic levels.

Cadmium can bind with the red blood cells. That's the main reason behind accumulation of cadmium in the kidneys and liver tissues. It has a very long biological half-life in any tissues [19]. It causes, "Itai-Itai" disease which was first identified among Japanese women due to the overexposure to this metal.

5.3.3 Chromium

The symbol of the Chromium is Cr. The atomic number is 24 with atomic weight and density of 51.99u and 7.19 g/cm^3, respectively. The Cr is stable element because its outermost shell is half-filled due to the transfer of electrons from ground state (4s) to highest state (3d). The electronic configuration of Cr is [Ar] 4s^1 3d^4.

The stable ionic species of Chromium are Cr (+3) and Cr (+6). Hexavalent Chromium (Cr^{+6}) in soil may naturally originating from chromite due to weathering process and is a carcinogen. But, Chromium concentration is increasing worldwide due to ore extraction, ferro-chrome alloy processing, hide tanning, and similar other industrial or metallurgical usage [24]. The concentration of Chromium in positively phototropic plant parts growing in natural soil types generally ranges between 20 and 200 µg/kg [19].

Elevated Chromium concentrations have been observed in soils and sediments of urbanized and industrialized areas because of vehicular emissions due to transportation, as well as surface runoff along with sewer, municipal, industrial overflows, and mining pollution [7, 9, 19, 24, 25]. Cr toxicity in plants shows symptoms like retardation of growth hormones, anomaly in root anatomy, curled and bleached leaves, chlorotic bands, and it impacts agricultural crops by reducing their yield [4, 19, 22].

Chromium do not easily translocated in higher tissues like shoot or leaves and hence is concentrated mainly in roots as it binds with the cell wall [26]. Applications of farm yard manures declines the Cr mobility. Hexavalent Cr is transformed to trivalent form in macrophyte cells and eventually disrupts the DNA replication/repair mechanism and the synthesized proteins. Brassicaceae family members are generally known to bioaccumulate elevated concentrations of Cr [22].

5.3.4 Copper

The scientific name of the Copper is cuprous and its symbol is Cu. It has an atomic number, atomic weight and density of 29, 63.546u, 8.96 g/cm^3, respectively. The electronic configuration of Copper is [Ar] 4s^1 3d^{10}. Cu is a d block element having fully filled outer most shell due to the sharing of electron with 3d orbital. That's the reason behind its apparent stability.

The concentration of Cu varies between 2 and 50 mg/kg in natural soil. Cu mainly occur in divalent state (Cu^{+2}). Basaltic igneous rocks and their weathered products have larger concentration of Cu [19].

In the tissue of the roots of macrophytes, Cu is present almost entirely in complexed form whereas this metal enters the root system as discrete ions. Root is observed to possess strong capability to bioaccumulate Cu in its cells while reducing its translocation to higher tissues like shoot or leaves. Hence, Cu deficiency symptoms can be observed in higher tissues of the plants. Cu is an essential micronutrition for the plants. Concentration of some enzymes like superoxidase and peroxidase is observed in leaf tissues under elevated Cu concentrations and may result in oxidative stress in the plants [22]. Elevated soil Cu concentration can cause toxicity in producer trophic level (macrophytes) and eventually can impact the ecosystem. Cu deficiency in plants can result in necrosis of the tissues [27].

5.3.5 Lead

The symbol of Lead is Pb. It has an atomic number, atomic mass, and density of 82, 207u, 11.34 g/cm^3, respectively. Average Lead concentration in continental sial (Silica-Alumina) layer is around 14.8 mg/kg. Pb exists mostly in the oxidation state of Pb^{+2} [19]. The geochemistry of Pb^{+2} is similar to the divalent group II metals and hence this post transition metal can easily replace K, Ba, Sr, and Ca [22].

The main toxicity of Lead in humans and animals is due to its impact on the central as well as peripheral nervous system. The symptoms due to Pb toxicity in human are represented by neurological impairments like decreased attention level, dementia, along with muscular tremor that can eventually result in 'Plumbism'. In macrophytes, it directly affects the morphology, growth, and photosynthetic process [28].

The Pb uptake by plants is dependent on soil properties like the capacity of the soil to exchange cations with the plants, Total Organic Matter (TOM) and acid-base balance. Subcellular effects of Pb toxicity on plant tissue result in depressed respiratory cycles (mostly Krebs cycle) and dark reactions of photosynthesis due to its impact on the Electron Transfer Chain (ETC) pathways involving the oxidation and reduction of Fe ions [22].

5.3.6 Nickel

The symbol of Nickel is Ni. The atomic number of Nickel is 28. Its atomic mass and density are 58.71u and 8.9 gm/cm^3, respectively. Nickel exists generally in the stable oxidation state of 0 and +2. It is found in almost all types of rocks and in the molten core. It is also found as residue in fossil fuels like coal and oil. In an uncontaminated soil, Ni concentration in the aboveground tissue generally ranges between 5 and 10 mg/kg [29].

Nickel plays very important role in organisms, and in case of its deficiency or toxicity, it shows variety of symptoms. Nickel toxic response is due to the metals affinity to replace other essential metals and to bind with proteins or enzymes having Nitrogen or Sulphur residues. The toxicity of Ni results due to its ability to replace other metals ions in enzymes and proteins or due to its binding to cellular enzymes containing S and N atoms [19, 22].

5.3.7 Zinc

The symbol of Zinc is Zn. The atomic number of Zn is 30. The atomic mass and density of Zn are 65.38 and 7.14 g/cm^3, respectively. It is found in all types of soil with concentration ranging between 10 and 100 mg/kg. It generally exists in the oxidation state of +2.

Zn is an essential nutrient. It is an essential co-factor for several enzymes in the body like carboxylase, dehydrogenases pepetidase, etc. It can help in metabolism of carbohydrates, proteins, and phosphates and also help in formation of auxins, Ribonucleic Acids (RNA), and proteins [22]. The Zinc concentration in plant should be around 60 mg/kg. But, higher Zn concentration can also results in plant toxicity. The elevated concentration of Zn can result in toxic responses in plants as well as benthic fauna. When the concentration is increased to 100–500 mg/kg, it results in 10% reduction in yield of plants and stunted root growth [19]. Zinc generally do not cause any chronic poisoning in human and wildlife due to the immobilization of this metal in the plant roots (producers) and thus effectively limiting it's transfer to the different trophic levels of the food-chain.

5.4 Metal Pollution in Soil Across the Globe

Metal pollution has a multitude of impact on ecosystem. The sources can be varied ranging from agricultural runoff, industrial effluent discharge, land conversion, aquaculture, mining, vehicular movements, to mention few. The metal pollution profile of soil has been given in Table 5.1.

It is evident that heavy metal contamination in the soil can rise due to anthropogenic impacts and concentrations differ from different regions across the globe as evident from the above case studies. As soil is of heterogeneous consistency, it is difficult to have a single permissible value for metal concentrations for different soil types. Again, understanding the anthropogenic interventions, chemical composition of parent rock material is pivotal in ascertaining the metal concentration in different edaphic

Table 5.1 Metal concentrations (mg/kg) in different soil types across the globe.

Site ID	Country	Types of soil	As	Cd	Cr	Cu	Pb	Ni	Zn	Reference
S1	Iran	Street dust	-	0.34	35.1	239	71.3	42.4	573	[33]
S2	China	Road dust	-	1.825	106.25	95	43		87.5	[34]
S3	Beijing China	Rural industrial soil	7.46	0.11	86.38	31.60	31.41	34.93	61.13	[35]
S4	Pakistan	Agricultural soil	-	2	-	-	14	-	-	[36]
S5	Argentina	Sediment soil		1.3	24	18.8	21.3		52.2	[37]
S6	China	Shanxi, Provincial administrative region	12.2	0.266	79	29.6	29.2	34.9	81.8	[38]
S7	China	Mines area soil	89.029	0.472			193.133			[39]
S8	Russia	Podzolic soil	-	-	20	30	4	60	50	[40]
S9	New Zealand	Top soil	3.5	0.097	14.1	12.0	11.3	8	49.6	[41]
S10	China	Hainan provincial administrative region	6.5	0.085	68.1	20.9	49.2	33.3	69.2	[38]

(Continued)

Table 5.1 Metal concentrations (mg/kg) in different soil types across the globe. (*Continued*)

Site ID	Country	Types of soil	As	Cd	Cr	Cu	Pb	Ni	Zn	Reference
S11	Pakistan	Mixed vegetation soil			81	45	90		255	[42]
S12	Portugal	Iberian soil		1.1	28.3	132.9	270.8	23.7	165.5	[43]
S13	Beijing China	Playground soil	0.64		0.59	0.17	0.07	0.70		[44]
S14	Azerbaijan	Sedimentary mixed soil at Oil field area	4.16	0.18	19.9	37.1	29.2	-	47.9	[45]
S15	China	-	16	-	-	46.4	252		286	[46]
S16	Pakistan	Roadside soil	-	0.09	-	-	0.09	0.76	0.04	[47]
S17	Vietnam	Agricultural soil around chromite mine	-	-	5750	-	-	5590	-	[30]
S18	China	-	9.37	1.51	66.7	63.7	30.8	10.1	325	[48]
S19	India	Sukinda chromite Mine, Odisha	-	-	401	-	67	45	42	[49]

(*Continued*)

Table 5.1 Metal concentrations (mg/kg) in different soil types across the globe. (*Continued*)

Site ID	Country	Types of soil	As	Cd	Cr	Cu	Pb	Ni	Zn	Reference
S20	China	Road dust	10.1	0.50	112	0.15	71.8	-	186	[50]
S21	Bhatin	Agricultural soil	2.35	0.62	1.84	3.53	2.68	1.16	1.22	[51]
S22	Pakistan	Soil of Northern province		1.0	202		8.0	58	17	[7]
S23	China	Huainan City, Agricultural soil in vicinity of coal mine	5.32		47.7	17.9	14.5		13.4	[52]
S24	Southeast China	Agricultural soil In peri-urban area	29.55	1.24	76.19	31.77	182.67	39.78	96.8	[53]
S25	Ahvaz Iran	Surface soil	6.91	5.58	107.42	187.06	189.14		93.56	[31]
S26	Australia	Surface soil	3	0.07	42	17	11	17	38	[54]
S27	Pakistan	Road soil	15	2	2.3	4			2.1	[55]
S28	China	Cropping soil of Wuhan		4.02	15.50	42.91	34.02	50.28	94.60	[32]
S29	India	Mine Soil		5.25	21430.01		203.2	7614.56	269.66	[9]

Table 5.2 Ecological Risk Index (ERI) of different study areas.

Site ID	Ecological Risk Factor (ERF)					Ecological Risk Index
	Cd	**Cr**	**Pb**	**Ni**	**Zn**	
S1	34	0.78	17.83	3.12	6.03	61.75
S2	182.5	2.36	10.75	0.00	0.92	196.53
S3	11	1.92	7.85	2.57	0.64	23.98
S4	200	0.00	3.5	0.00	0.00	203.50
S5	130	0.53	5.325	0.00	0.55	136.41
S6	26.6	1.8	7.3	2.6	0.9	39.1
S7	47.2	0.00	48.28	0.00	0.00	95.48
S8	0	0.44	1	4.41	0.53	6.38
S9	9.7	0.31	2.825	0.59	0.52	13.95
S10	8.50	1.51	12.30	2.45	0.73	25.49
S11	0	1.80	22.5	0.00	2.68	26.98
S12	110	0.63	67.7	1.74	1.74	181.81
S13	0	0.01	0.018	0.05	0.00	0.08
S14	18	0.44	7.3	0.00	0.50	26.25
S15	0	0.00	63	0.00	3.01	66.01
S16	9	0.00	0.023	0.06	0.00	9.08
S17	0	127.78	0	411.03	0.00	**538.81**
S18	151	1.48	7.7	0.74	3.42	164.35
S19	0	8.91	16.75	3.31	0.44	29.41
S20	50	2.49	17.95	0.00	1.96	72.40
S21	62	0.04	0.67	0.09	0.01	62.81
S22	100	4.49	2	4.26	0.18	110.93
S23	0	1.06	3.625	0.00	0.14	4.83

(Continued)

Table 5.2 Ecological Risk Index (ERI) of different study areas. (*Continued*)

Site ID	Ecological Risk Factor (ERF)					Ecological Risk Index
	Cd	Cr	Pb	Ni	Zn	
S24	124	1.69	45.6675	2.93	1.02	175.30
S25	558	2.39	47.285	0.00	0.98	***608.66***
S26	7	0.93	2.75	1.25	0.40	12.33
S27	200	0.05	0	0.00	0.02	200.07
S28	402	0.34	8.505	3.70	1.00	**415.54**
S29	525	476.2	50.8	559.89	2.84	***1614.75***

The ERI values falling in the category of "Considerable Ecological Risk" is marked in **bold**, while the values that is in the category of "Very High Ecological Risk" has been indicated in *Italics* and put in **bold** font for easy understanding.

conditions. For example, works of Naz *et al.* (2018) indicated a high concentration of Cr in roadside soil, agricultural soil, along with mining overburden dump anthrosol near a Chromite mine in India [9]. This exemplifies the impact of mining activity in most of the soil that has increased the concentrations of Cr.

The ERI data in Table 5.2 indicates that the agricultural soil near the Chromite mine of Vietnam, the traffic and industrial pollution in the Wuhan province of China, and industrial and vehicular traffic affecting the dust pollution at the street side dust/loose soil in Ahvaz region of Iran has pollution in the category of "Considerable Ecological Risk" [9, 30–32]. The mining soil around the Sukinda Chromite Mines investigated by Naz *et al.* 2018 has pollution in the range of "very high ecological risk" due to the impact of Chromite mine waste [9].

5.5 Ecological and Human Health Risk Impacts of Heavy Metals

Heavy metals stay in the ecosystem for indefinite period of time. Generally, it comes into the food chain through bioaccumulation and translocation process in producers which are the green plants. Heavy metals can exist in different forms in the soil but only their bioavailable portion is bioconcentrated in the root [4–6, 8–10].

Bioaccumulation: It can be defined as progressive collection of substances in tissues, cell, or organs.

Bioconcentration: How many times the concentration of a particular substance, in this case the heavy metal, is more in the tissues of living organism in comparison with the environment of their habitat that can be water or soil.

Translocation: Generally, it is equated as the Translocation Factor and it can be defined as the concentration of substances in higher tissues of plants like shoot, leaves in compared with the concentration in root.

Biomagnification: It an be defined as increase in concentration of substances, in this case heavy metals, and in different trophic levels of the food chain.

Hence, as discussed in Figure 5.1, metal pollutants get biomagnified in the higher trophic levels through the food chain. In the figure, the pollution sources spew metals into the environment which gets biomagnified in different trophic levels while it can also result in human health risk via consumption pathway.

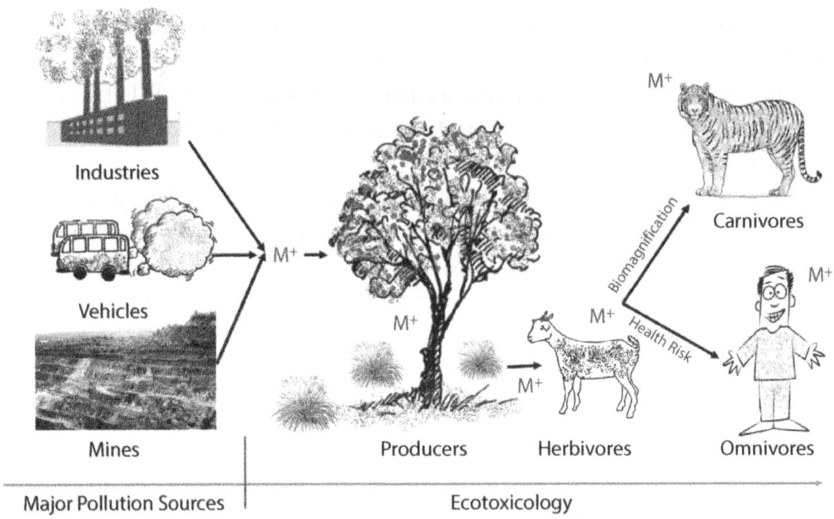

Figure 5.1 Ecotoxicological and Human Health impact pathway of Toxic metals. The pollution sources of the heavy metals (M^+) is given on the left hand side and due to the contamination the metal get bioaccumulated in the different trophic levels of the simple food chain comprising of producers (Green macrophytes), obligatory herbivores, and the topmost of the trophic level, i.e., the Carnivores. Humans are omnivores that consume both plant and animals for nutrition and heavy metals can get biomagnified in their tissues causing health hazards. Topmost predator or Carnivores also tend to have biomagnified concentrations of the heavy metals in their body due to their bottom-up transfer in the food chain.

5.6 Conclusion

This book chapter dealt with the sources, extent of pollution and geochemistry of Cadmium, Arsenic, Chromium, Lead, Copper, Zinc, and Nickel. Cd, As, Cr, and Pb toxicity are a major concern in the ecosystem and to human health because of their potential to be biomagnified in higher trophic levels as well as impairing the enzymatic metabolic pathways in living cells. Hence, understanding the geochemistry of these elements in soil is important in the context of pollution estimation and remediation. As there are no set guidelines for the extent of pollution in different soil types and different parent rock material result in different metal concentrations, an index-based approach has been used in this chapter to analyze the metal pollution levels with their probabilistic impact on the ecosystem. It is concluded that mining activity and industrial waste disposal pose the maximum threat to the ecosystem by mobilizing toxic persistent metals and necessary actions or management plans need to be taken to address this pressing issue of global concern.

References

1. Sarwar, N., Imran, M., Shaheen, M.R., Ishaq, W., Kamran, A., Matloob, A., Rehim, A., Hussain, S., Phytoremediation strategies for soils contaminated with heavy metals: Modifications and future perspectives. *Chemosphere*, 171, 710–721, 2017.
2. Sarwar, N., Malhi, S.S., Zia, M.H., Naeem, A., Bibi, S., Farid, G., Role of mineral nutrition in minimizing cadmium accumulation by plants. *J. Sci. Food Agric.*, 90, 6, 925–937, 2010.
3. Usman, A.R.A. and Mohamed, H.M., Effect of microbial inoculation and EDTA on the uptake and translocation of heavy metal by corn and sunflower. *Chemosphere*, 76, 7, 893–899, 2009.
4. Chowdhury, A. and Maiti, S.K., Identification of metal tolerant plant species in mangrove ecosystem by using community study and multivariate analysis: a case study from Indian Sunderban. *Environ. Earth Sci.*, 75, 9, 744, 2016.
5. Chowdhury, A. and Maiti, S.K., Assessing the ecological health risk in a conserved mangrove ecosystem due to heavy metal pollution: a case study from Sundarbans biosphere reserve, India. *Hum. Ecol. Risk Assess.*, 22, 7, 1519–1541, 2016.
6. Kumari, P., Chowdhury, A., Maiti, S.K., Assessment of heavy metal in the water, sediment, and two edible fish species of Jamshedpur Urban Agglomeration, India with special emphasis on human health risk. *Hum. Ecol. Risk Assess.*, 24, 6, 1477–1500, 2018.

7. Nawab, J., Khan, S., Shah, M.T., Gul, N., Ali, A., Khan, K., Huang, Q., Heavy metal bioaccumulation in native plants in chromite impacted sites: A search for effective remediating plant species. *Clean-Soil Air Water*, 44, 1, 37–46, 2016.
8. Naz, A., Chowdhury, A., Mishra, B.K., Gupta, S.K., Metal pollution in water environment and the associated human health risk from drinking water: A case study of Sukinda chromite mine, India. *Hum. Ecol. Risk Assess.*, 22, 7, 1433–1455, 2016.
9. Naz, A., Chowdhury, A., Mishra, B.K., Karthikeyan, K., Distribution of heavy metals and associated human health risk in mine, agricultural and roadside soils at the largest chromite mine of India. *Geochem. Health*, 40, 2155–2175, 2018.
10. Naz, A., Mishra, B.K., Gupta, S.K., Human health risk assessment of chromium in drinking water: A case study of sukinda chromite mine, Odisha, India. *Expo. Health*, 8, 2, 253–264, 2016.
11. Fernández, S., Poschenrieder, C., Marcenò, C., Gallegod, J.R., Jiménez-Gámezd, D., Buenoc, A., Afifd, E., Phytoremediation capability of native plant species living on Pb-Zn and Hg-As mining wastes in the Cantabrian range, north of Spain. *J. Geochem. Explor.*, 174, 10–20, 2017.
12. Chowdhury, A., Naz, A., Maiti, S.K., Health risk assessment of 'tiger prawn seed' collectors exposed to heavy metal pollution in the conserved mangrove forest of Indian Sundarbans: A socio-environmental perspective. *Hum. Ecol. Risk Assess.*, 23, 2, 203–224, 2017.
13. Raj, D., Chowdhury, A., Maiti, S.K., Ecological risk assessment of mercury and other heavy metals in soils of coal mining area: a case study from Eastern part of Jharia coal field, India. *Hum. Ecol. Risk Assess.*, 23, 4, 1767–787, 2017.
14. Du, C., Wu, J., Bashir, M.H., Shaukat, M., Ali, S., Heavy metals transported through a multi-trophic food chain influence the energy metabolism and immune responses of Cryptolaemus montrouzieri. *Ecotoxicology*, 28, 4, 422–428, 2019.
15. Ganeshkumar, A., Arun, G., Vinothkumar, S., Rajaram, R., Bioaccumulation and translocation efficacy of heavy metals by Rhizophora mucronata from tropical mangrove ecosystem, Southeast coast of India. *Ecohydrol. Hydrobiol.*, 19, 1, 66–74, 2019.
16. Turekin, K.K. and Wedepohl, K.H., Distribution of the elements in some major units of the earth's crust. *Geol. Soc. Am. Bull.*, 72, 2, 175–192, 1961.
17. Muller, G., Heavy Metals in the Sediment of the Rhine-Changes Seity. *Umsch. Wiss. Tech.*, 79, 778–783, 1979.
18. Hakanson, L., An ecological risk index for aquatic pollution control: A sedimentological approach. *Water Res.*, 14, 8, 975–1001, 1980.
19. Alloway, B.J. (Ed.), *Heavy metals in soils: trace metals and metalloids in soils and their bioavailability*, vol. 22, Springer Science & Business Media, Dordrecht Heidelberg New York London, 2012.

20. Banerjee, S., Datta, S., Chattyopadhyay, D., Sarkar, P., Arsenic accumulating and transforming bacteria isolated from contaminated soil for potential use in bioremediation. *J. Environ. Sci. Health Part A*, 46, 14, 1736–1747, 2011.
21. Pongratz, R., Arsenic speciation in environmental samples of contaminated soil. *Sci. Total Environ.*, 224, 1-3, 133–141, 1998.
22. Kabata-Pendias, A. and Mukherjee, A.B., *Trace elements from soil to human*, Springer Science & Business Media, Berlin Heidelberg New York, 2007.
23. Vig, K., Megharaj, M., Sethunathan, N., Naidu, R., Bioavailability and toxicity of cadmium to microorganisms and their activities in soil: A review. *Adv. Environ. Res.*, 8, 1, 121–135, 2003.
24. Martello, L., Fuchsman, P., Sorensen, M., Magar, V., Wenning, R.J., Chromium geochemistry and bioaccumulation in sediments from the lower Hackensack River, New Jersey. *Arch. Environ. Con. Tox.*, 53, 3, 337–350, 2007.
25. Paul, J.F., Comeleo, R.L., Copeland, J., Landscape metrics and estuarine sediment contamination in the mid-Atlantic and southern New England regions. *J. Environ. Qual.*, 31, 3, 836–845, 2002.
26. Zayed, A., Lytle, C.M., Qian, J.H., Terry, N., Chromium accumulation, translocation and chemical speciation in vegetable crops. *Planta*, 206, 2, 293–299, 1998.
27. Nagajyoti, P.C., Lee, K.D., Sreekanth, T.V.M., Heavy metals, occurrence and toxicity for plants: A review. *Environ. Chem. Lett.*, 8, 3, 199–216, 2010.
28. Flora, G., Gupta, D., Tiwari, A., Toxicity of lead: A review with recent updates. *Interdiscip. Toxicol.*, 5, 2, 47–58, 2012.
29. Scott-Fordsmand, J.J., Toxicity of nickel to soil organisms in Denmark, in: *Rev Environ Contam T*, pp. 1–34, Springer, New York, NY, 1997.
30. Kien, C.N., Noi, N.V., Son, L.T., Ngoc, H.M., Tanaka, S., Nishina, T., Iwasaki, K., Heavy metal contamination of agricultural soils around a chromite mine in Vietnam. *J. Soil Sci. Plant Nutr.*, 56, 2, 344–356, 2010.
31. Ghanavati, N., Nazarpour, A., De Vivo, B., Ecological and human health risk assessment of toxic metals in street dusts and surface soils in Ahvaz, Iran. *Environ. Geochem. Health*, 41, 2, 875–891, 2019.
32. Yang, Y., Christakos, G., Guo, M., Xiao, L., Huang, W., Space-time quantitative source apportionment of soil heavy metal concentration increments. *Environ. Pollut.*, 223, 560–566, 2017.
33. Kamani, H., Mirzaei, N., Ghaderpoori, M., Bazrafshan, E., Rezaei, S., Mahvi, A.H., Concentration and ecological risk of heavy metal in street dusts of Eslamshahr, Iran. *Hum. Ecol. Risk Assess.*, 24, 4, 961–970, 2018.
34. Jiang, Y., Shi, L., Guang, A.L., Mu, Z., Zhan, H., Wu, Y., Contamination levels and human health risk assessment of toxic heavy metals in street dust in an industrial city in Northwest China. *Environ. Geochem. Health*, 40, 5, 2007–2020, 2018.
35. Jiang, Y., Chao, S., Liu, J., Yang, Y., Chen, Y., Zhang, A., Cao, H., Source apportionment and health risk assessment of heavy metals in soil for a township in Jiangsu Province, China. *Chemosphere*, 168, 1658–1668, 2017.

36. Rehman, Z.U., Khan, S., Brusseau, M.L., Shah, M.T., Lead and cadmium contamination and exposure risk assessment via consumption of vegetables grown in agricultural soils of five-selected regions of Pakistan. *Chemosphere*, 168, 1589–1596, 2017.
37. Castro, L.N., Rendina, A.E., Orgeira, M.J., Assessment of toxic metal contamination using a regional lithogenic geochemical background, Pampean area river basin, Argentina. *Sci. Total Environ.*, 627, 125–133, 2018.
38. Chen, H., Teng, Y., Lu, S., Wang, Y., Wang, J., Contamination features and health risk of soil heavy metals in China. *Sci. Total Environ.*, 512, 143–153, 2015.
39. Fan, Y., Zhu, T., Li, M., He, J., Huang, R., Heavy metal contamination in soil and brown rice and human health risk assessment near three mining areas in central China. *J. Healthc. Eng.*, 2017, 1–9, 2017.
40. Evseev, A.V. and Krasovskaya, T.M., Toxic metals in soils of the Russian North. *J. Geochem. Explor.*, 174, 128–131, 2017.
41. Martin, A.P., Turnbull, R.E., Rissmann, C.W., Rieger, P., Heavy metal and metalloid concentrations in soils under pasture of southern New Zealand. *Geoderma Reg.*, 11, 18–27, 2017.
42. Su, H., Guo, P., Zhang, Y., Deng, J., Wang, M., Sun, Y., Wu, Y., Effects of planting patterns on the concentration and bioavailability of heavy metals in soils during wetland restoration. *Int. J. Environ. Sci. Technol.*, 16, 2, 853–86, 2019.
43. Vareda, J.P., Valente, A.J., Durães, L., Heavy metals in Iberian soils: Removal by current adsorbents/amendments and prospective for aerogels. *Adv. Colloid Interface Sci.*, 237, 28–42, 2016.
44. Jin, Y., O'Connor, D., Ok, Y.S., Tsang, D.C., Liu, A., Hou, D., Assessment of sources of heavy metals in soil and dust at children's playgrounds in Beijing using GIS and multivariate statistical analysis. *Environ. Int.*, 124, 320–328, 2019.
45. Khalilova, H. and Mammadov, V., Assessing the Anthropogenic Impact on Heavy Metal Pollution of Soils and Sediments in Urban Areas of Azerbaijan's Oil Industrial Region. *Pol. J. Environ. Stud.*, 25, 1, 159–166, 2016.
46. Xiao, R., Wang, S., Li, R., Wang, J.J., Zhang, Z., Soil heavy metal contamination and health risks associated with artisanal gold mining in Tongguan, Shaanxi, China. *Ecotoxicol. Environ. Saf.*, 141, 17–24, 2017.
47. Khalid, N., Noman, A., Aqeel, M., Masood, A., Tufail, A., Phytoremediation potential of Xanthium strumarium for heavy metals contaminated soils at roadsides. *Int. J. Environ. Sci. Technol.*, 16, 4, 2091–2100, 2019.
48. Ding, F., He, Z., Liu, S., Zhang, S., Zhao, F., Li, Q., Stoffella, P.J., Heavy metals in composts of China: historical changes, regional variation, and potential impact on soil quality. *Environ. Sci. Pollut. Res.*, 24, 3, 3194–3209, 2017.
49. Pattnaik, B.K. and Equeenuddin, S.M., Potentially toxic metal contamination and enzyme activities in soil around chromite mines at Sukinda Ultramafic Complex, India. *J. Geochem. Explor.*, 168, 127–136, 2016.

50. Tang, Z., Huang, Q., Yang, Y., Nie, Z., Cheng, J., Yang, J., Chai, M., Polybrominated diphenyl ethers (PBDEs) and heavy metals in road dusts from a plastic waste recycling area in north China: Implications for human health. *Environ. Sci. Pollut. Res.*, 23, 1, 625–637, 2016.
51. Giri, S., Singh, A.K., Mahato, M.K., Metal contamination of agricultural soils in the copper mining areas of Singhbhum shear zone in India. *J. Earth Syst. Sci.*, 126, 4, 49, 2017.
52. Tang, Q., Li, L., Zhang, S., Zheng, L., Miao, C., Characterization of heavy metals in coal gangue-reclaimed soils from a coal mining area. *J. Geochem. Explor.*, 186, 1–11, 2018.
53. Huang, Y., Chen, Q., Deng, M., Japenga, J., Li, T., Yang, X., He, Z., Heavy metal pollution and health risk assessment of agricultural soils in a typical peri-urban area in southeast China. *J. Environ. Manage.*, 207, 159–168, 2018.
54. Abraham, J., Dowling, K., Florentine, S., Assessment of potentially toxic metal contamination in the soils of a legacy mine site in Central Victoria, Australia. *Chemosphere*, 192, 122–132, 2018.
55. Khan, Z., II, Ugulu, I., Umar, S., Ahmad, K., Mehmood, N., Ashfaq, A., Sohail, M., Potential toxic metal accumulation in soil, forage and blood plasma of buffaloes sampled from Jhang, Pakistan. *Bull. Environ. Contam. Toxicol.*, 101, 2, 235–242, 2018.

6

Isotope Geochemistry

Praveen Kumar Yadav*, Amit Kumar Mauraya, Chinky Kochar, Lakhan Taneja and S. Swarupa Tripathy

Academy of Science and Innovative Research (AcSIR), Chemical and Food BND Group, IRM (BND), CSIR-National Physical Laboratory, Dr K.S. Krishnan Marg, New Delhi, India

Abstract

Stable isotopes have shown very versatile applications in various fields. In case of geochemistry, two types of stable isotopes have been used, i.e., traditional stable isotopes including C, N, O, S, and H, and nontraditional stable isotopes such as Zn, Cu, Fe, Pd, Cd, and other metals. Though, traditional isotopes have been utilized from the beginning of the 20th century and hence shown very significant application in geochemistry. The traditional isotopes have been utilized in various geochemical applications. For example, traditional isotopes have been used as geothermometers, tracer in hydrological, biological systems, ore deposits, hydrothermal systems, etc. However, nontraditional isotopes have shown limited study due to lack of advanced technique and method. But, the recent advancement in the techniques and the development of various advanced method have increased their applications in various fields. The nontraditional isotopes have been used in various geochemical applications such as for tracing of toxic environmental pollutants like Cr and Hg, application in mining, etc.

Keywords: Stable isotopes, geochemistry, traditional isotopes, nontraditional isotopes, geothermometers, environmental pollutants

6.1 Introduction

The geochemistry of stable isotopes deals the variations in elemental isotopic compositions due to the physicochemical processes. The study on

**Corresponding author*: kpraveen.yadav414@gmail.com

Inamuddin, Mohd Imran Ahamed, Rajender Boddula and Tariq Altalhi (eds.) Geochemistry: Concepts and Applications, (93–110) © 2021 Scrivener Publishing LLC

stable isotope geochemistry started in early decades of the 20th century along with the evolution of modern physics. The main precursors of these discoveries were the exploration of variation in isotopic compositions of lighter elements along with the discovery of neutron in the first half of the 20th century. Earliest publications related stable isotope geochemistry were published by Harold Urey [1] and Bigeleisen *et al.* [2] entitled as "The Thermodynamic Properties of Isotopic Substances" and "Calculation of equilibrium constants for isotopic exchange reactions". Urey hypothesized that isotope fractionation could be predicted on the basis of theoretical grounds which could also provide valuable information related to geology. He further studied the isotopic compositions of natural substances and effect of temperature on isotope fractionation.

Earlier, the traditional isotopes such as H, C, N, O, and S were the main elements of interest in the study of stable isotope geochemistry. Later on, B and Li were also studied excessively in isotope geochemistry. Significant data on sources and processes which have impacted the historical backdrop of the samples can be found from Stable isotope ratios of elements. Additionally, the traditional isotopes have been effectively utilized in environmental geochemistry for several decades to examine the environmental cycling of elements [3–5]. These supposed "conventional" stable isotope frameworks incorporate just components that can be advantageously changed into a vaporous structure and examined by gas-source mass spectrometers [6]. However, in case of natural variation study of heavier isotope, lack of suitable techniques has limited their application as well as further advancement for long time. Though, recent advancement in the techniques offers higher precision measurement and highly sensitive instruments like MC-ICP-MS, which enhanced the field of application by exploring more elements such as Ti, Si, Mg, Ca, Cl, Cr, Fe, Zn, Cu, Tl, U, Mo, Ge, and Se. Small isotopic variation have been observed in these metals which has provided very useful geological information and the geochemical exploration of these metals is still continue [7].

6.2 Basic Definitions

6.2.1 The Notation

The variations in stable isotope ratios are generally expressed as per mil deviations, δ, from some standard and are normally in the range of parts per thousand to parts per hundred [8]. For instance, isotope ratios of O are expressed as per mil deviations from standard as follows:

$$\delta^{18}O = (^{18}O/^{16}O)_{sample}/(^{18}O/^{16}O)_{standard} - 1 \quad (6.1)$$

6.2.2 The Fractionation Factor

The fractionation factor, α, is represented as follows:

$$\alpha_{A\text{-}B} = R_A/R_B \quad (6.2)$$

where R_A represent the isotope ratio of one phase and R_B for another phase. The isotope fractionation between two phases can also be represented as $\Delta_{A\text{-}B} = \delta_A - \delta_B$. The relationship between Δ and α is

$$\Delta \approx (\alpha - 1)10^3 \quad (6.3)$$

Relationship between and equilibrium constant of thermodynamics is

$$\alpha_{A\text{-}B} = (K/K_\infty)^{1/n} \quad (6.4)$$

where K is the equilibrium constant, K_∞ is the equilibrium constant at infinite temperature, and *n* is the number of atoms exchanged.

6.2.3 Isotope Fractionation

Both equilibrium and kinetic effects can produce fractionation in isotopes. Kinetic effects might be naturally expected, but the equilibrium effects may be surprising, initially. It is well known that the electronic structure directs the chemical properties of an element, and there is no role of the nucleus in chemical interactions. In any reaction, a shift in the isotope proportion among the reactant and product is caused by stable isotope fractionation. However, this change is almost negligible; hence, no change appears in isotopic mass balance of the whole system. For instance, if one reservoir is showing isotopic enrichment, another must be balanced by corresponding depletion of isotopes. However, in small reservoir, a significant isotopic variation can be encouraged easily but in large reservoir, the small isotopic variation is undetectable.

6.2.3.1 Kinetic Isotope Fractionation

The variation in rate of reaction of lighter and heavier isotopes and incomplete reactions are the major source of kinetic isotope effects. This effect also arises due to several other processes such as biological, diffusion, and evaporation processes where enzyme is involved during the bond breaking process which induces the breaking of lighter isotope bonds owing to their higher zero-point energies [6, 10]. The extent of reaction determines the impact of kinetic effects on isotopic composition of any sample. When reactions proceed far toward completion, light isotopes are continuously removed preferentially; therefore, the largest effects are observed in remaining reactants which are strongly enriched in heavy isotopes. It is obvious that if reactants are totally changed into product, no sign of kinetic effects will be there which ultimately led to the similar isotopic composition in reactant and product. The development of isotope ratio during the any unidirectional and incomplete reactions within closed system is mostly studied by using the Rayleigh fractionation model. The Rayleigh model can be applied in various systems, for example, kinetically controlled systems with unidirectional rate of reaction as well as all other systems where the physical removal of product is involved etc. These models can also be utilized for the quantification of degree of transformation or to find out the fractionation factor of any particular process using the data of isotopic analysis. However, these models face certain limitations also, for example, these models can be used in homogenous reactant pools only, which allow for the suitable fractionation of isotopes in reactant pool.

6.2.3.2 Equilibrium Isotope Fractionation

When two phases react with forward and backward reactions taking place at equal rates then they show equilibrium isotope effect. Here, the difference in energy of bonding environment of reactants (which reached at isotopic equilibrium) controls the relative abundance of isotopes. Though, concentration and isotopic equilibrium both needed different time scale to reach the equilibrium [7]. The difference in bonding environment of heavy and light isotopes can be seen at isotopic equilibrium where the heavier isotopes form lower energy bond, i.e., stronger bond whereas the lighter isotope forms weaker bond at equilibrium condition [9]. With increase in temperature, equilibrium isotope effect decreases, and this is the reason for their application as paleothermometer.

Equilibrium isotope effects were also calculated by using molecular modeling for various elements and between the elements as well (e.g., oxidation states and complexes) [11]. Metal isotope fractionation can be easily understood from theoretical point of view using computational methods. These methods were also utilized to find out the determine extent of equilibrium effects and the direction in particular systems [10]. It is a rule of thumb that, in redox reactions, the oxidized species is enriched with the heavy isotopes, and hence, the remaining lighter isotopes enriches in reduced form. Another thumb of rule is that the heavier isotope enrichment tendency is observed in metal species having short bond length as well as low coordination number owing to low zero-point energy. However, precaution should be taken before the application of these rules. Because, the equilibrium effects mainly depend on the difference in the vibrational frequencies, i.e., stiffness of bond, it does not depend on the strength of bond. For example, in compounds (HgS), Hg–S bond is stronger bond but it showed lighter isotope enrichment because this compound shows low stiffness and "soft-acid soft-base" interaction. While, in another compound, i.e., HgO, Hg–O bond is weak but with high stiffness and hence shows enrichment of heavier isotope [12]. Additionally, in comparison to kinetic effects, equilibrium effects do not depend on variables (e.g., conditions, rates, and processes) [10], usually assessed from theoretical calculations.

6.2.4 Mass Dependent and Independent Fractionations

S and O both elements have greater than two isotopes similar to many other elements. So, how can we assume the fractionations between various isotope ratios and the relationship among them? Let's discuss with an example. Suppose, any particular sample showed 4‰ fractionation of $\delta^{18}O$, what value we can predict for $\delta^{17}O$? So, our first speculation will be that the fractionation difference should be proportional to mass difference, i.e., the value for $\delta^{17}O$ should be at least 2‰. M. W. White [13] studied the fractionation in detail which showed the complex relationship between fractionation factors and mass, i.e., fractionation is complex function of mass (from eq. 9.26 of chapter 9 by M. W. White). The mass given in eq. 9.26 is a reduced mass ($m^{3/2}$) and an exponential term. Therefore, the fractionation value for $\delta^{17}O$ would be around 0.52 and it would not be exactly 0.50 to that of $\delta^{18}O$ value. Consequently, the fractionation, in this example, is proportional to the mass difference hence it is called the mass-dependent fractionation (MDF). On the other hand, where the value of fractionation of $\delta^{17}O$ is close to one that to $\delta^{17}O$ value, then it is called the mass-independent fractionation (MIF) because the fractionation is independent to the mass difference. MIF is rare and the first MIF was seen in O isotope of meteorites sample. Later on, it has also been reported in O isotopes of atmospheric gases as well as in sulfur isotope of sulfur-loaded aerosols in ice and Archean sediments [14, 15].

6.3 Application of Traditional Isotopes in Geochemistry

6.3.1 Geothermometer

The traditional stable isotopes have shown very versatile applications, and application as geothermometer is one of them. The geothermometers of stable isotope are based on the equilibrium constant which depend on temperature which is expressed as

$$\ln K = \ln \alpha = A + B/T^2 \quad (6.5)$$

where A and B are constants which is the function of temperature and K is equilibrium constant which become 0 at absolute zero temperature and show complete isotope separation whereas at infinite temperature equilibrium constant tend to be unity with no isotopic separation. Reason behind this question is entropy because on increasing the temperature the entropy of the system increases which also causes the random mixing of isotopes between phases at infinite temperature. While, at absolute zero temperature entropy is minimum, so no disorder and no mixing of isotopes takes place as well. So, according to the rule, the temperature can also be calculated from the isotope fractionation between any two phases. There are number of isotopes geothermometers [16] revealing the temperature dependence of fractionation factors between the different minerals. Due to temperature dependence of equilibrium constant, most of the isotope geothermometers are used at lower temperature because at higher temperature fractionation is very less, and hence, the temperature measurement with accuracy becomes difficult. There are several advantages of Isotope geothermometers over the chemical geothermometers. For example, in isotope geothermometers, there is no effect of volume and almost negligible effect of pressure on isotope exchange reaction and equilibrium constant, respectively. Another benefit is that the isotope geothermometers can be used in pure phases also unlike the conventional geothermometers which is, generally, depend on solid solution. Effect of composition is seen on the element which is involved in isotope exchange.

6.3.2 Isotopes in Biological System

Isotope fractionation in biological processes is very complex and occur largely in comparison to other medium. C, N, and S are the most prominent elements of biological system and hence, show the isotope fractionation in them due to biological processes. It has been noticed that the maximum fractionation occurred during the production of organic matter by primary producers, in case of C isotope. Mainly, the organic matter is produced during the photosynthesis and somewhat by chemosynthesis (at mid-ocean ridge hydrothermal vents) as well. Additionally, the small number of fractionations are also observed through subsequent reactions by heterotrophs which consume the primary producers.

6.3.2.1 Carbon (C)

The main cause of isotope fractionation in C is various biological processes involved in biological systems and photosynthesis is major contributor among all biological processes for C isotope fractionation, in case of plant body. The fractionation occur during the photosynthesis is, basically, due to kinetic effect. The study of Park and Epstein [17, 18] revealed the several steps of fractionation during the photosynthesis in plants. In terrestrial plants, isotope fractionation of −4.4‰ was observed during the diffusion of CO_2 (i.e., the first step) into the leaf, through the stomata. Whereas, in case of aquatic plants which uses these dissolved CO_2 or HCO_3^- for photosynthesis:

$$CO_{2(g)} \rightarrow CO_{2(aq)} + H_2O \rightarrow H_2CO_3 \rightarrow H^+ + HCO_3^- \qquad (6.6)$$

The dissolution of CO_2 revealed the equilibrium fractionation of +0.9 ‰ whereas the hydration and dissociation of H_2CO_3 showed +7.0‰ to +8‰, equilibrium fractionation. Additionally, all C_3 plants utilize the ribulose bisphosphate carboxylase oxygenase (RUBISCO) enzyme during the photosynthesis where the CO_2 is converted into 3- phosphoglyceric acid (also called carboxylation) which further converted into carbohydrate with the regeneration of ribulose bisphosphate. So, in higher terrestrial plants, kinetic fractionation of −29.4‰ was observed during the carboxylation process. While, in case of bacterial carboxylation, the fractionation was smaller, i.e., −20‰. Thus, for terrestrial plants a fractionation of about −34‰ is expected from the sum of the individual fractionations. The actual observed total fractionation is in the range of −20‰ to −30‰. A general assumption regarding the fractionation is that the extent of fractionation in C isotopes depends on the ratio of internal and external environmental concentration of CO_2, due to the difference between the expected and observed fractionation value [18]. After the process of photosynthesis, almost all organic matter produced during the photosynthesis get converted into other organic compounds utilized in different functions of body of organisms. Thus, isotope fractionation further happens during the conversion of organic compounds which was noticed as kinetic fractionation arises due to organic C-H bonds. The organic matter also produced through the chemosynthesis process where the fixation of CO_2 take place through chemical reaction unlike photosynthesis. Therefore, the isotope fractionation also occurs due to the chemosynthesis, and both chemosynthesis and photosynthesis show signature of large isotope fractionation in carbon isotope.

6.3.2.2 *Nitrogen (N)*

Another most crucial biological element is N which is essential element of proteins, amino acids, and other components like DNA and RNA. Biological assimilation of nitrogen led to the kinetic isotope fractionation. Terrestrial plants use nitrogen in the form of ammonia and N_2 fixation occurs with help of symbiotic bacteria. Several other plants such as marine algae and blue green algae use nitrogen either its oxidized form, e.g., NO_3^- and NO_2^- or directly as N_2. So, the assimilation of nitrogen involves several steps like with the carbon, and hence, the fractionation also occurs in nitrogen isotope during the assimilation process. Whereas, in denitrification process bacteria reduce the nitrate without its assimilation. During the denitrification process substantial kinetic fractionation occur which led to the enrichment of ^{15}N by 6‰ to 7‰ in residual nitrate. The planktons either nitrogen fixing or non-nitrogen fixing have shown isotopic composition of nitrogen ranging −3‰ to +12‰ [19].

6.3.3 Isotopes in Archaeology

Different foodstuffs show different isotopic composition of C and N which finally get accumulated into the bone collagen. So, the isotopic ratio measurement of C and N in bone reveals the dietary behavior of ancient humans. For example, isotopic analysis of bone collagen from human remains [20, 21] exhibited that around 4000 BC, Tehuacan Valley dwellers were heavily dependent on maize, while the archaeological surveys revealed that no use of maize until 1500 BC. Additionally, the dietary behavior of ancient people can also find out by studying the burned vegetable and meshed grain onto the potsherds. For instance, a study by DeNiro and Hasdorf [19] revealed that isotopic analysis of vegetables which were treated in similar condition as burials in soil and showed the large shift in $\delta^{13}C$ and $\delta^{15}N$ value. Whereas, when the vegetables were charred, it did not show any variation in isotopic composition of C and N close to the isotopic compositions of burials. However, burning of vegetable showed small fractionation of about 2‰ or 3‰, which is of very little significance.

6.3.4 Isotopes in Fossils and the Earliest Life

As large isotope fractionation noticed in C isotope of biological compounds, similarly, large isotope fractionation, i.e., −26‰ observed in $\delta^{13}C$ in carbonate rocks of West Greenland [22]. Later on, another study [23] reported fractionation ranges from −20 to −50‰ in $\delta^{13}C$ of grains of apatite contained graphite of same place. Rosing [24] examined graphite in pelagic and tubiditic metasedimentary rocks of Isua greenstone belt and reported the fractionation of −19‰ in $\delta^{13}C$. Additionally, the negative values of $\delta^{13}C$ were used as indication of a biogenic origin of the C, and hence, as indicator of life on earth at that time. But, this technique remained controversial. All controversies are related to the enormously complex geological history of the area. Because most of the rocks are metamorphosed and highly deformed, and the structure of precursor rocks is very difficult to interpret.

6.3.5 Isotopes in Hydrothermal and Ore Deposits

Entrance of large igneous bodies into the crust led to the heating of surrounding rock as well as the water present in their pores and cracks. So, the water available in these hydrothermal systems reacts with the hot rock, and hence, the exchange of isotopes takes place between rocks and water. Craig *et al.* [25] demonstrated the different type of hydrothermal systems and revealed that water of these systems was meteoric (lake, river, snow, rain, and water) rather than the magmatic (magma water). In chloride type geothermal waters, the value of δ D is similar to the ground water and local precipitation water, while the value of $\delta^{18}O$ do not follow this trend. In geothermal waters, the value of $\delta^{18}O$ shifted toward higher value which may be due to the reaction between hot rock and local meteoric water at high temperature (~<300°C). However, in case of rocks, very small change in isotopic composition of hydrogen due to the almost negligible availability of hydrogen in rocks. Additionally, the hydrothermal systems with sulfur-rich and/or acidic waters have different δ D value than the local meteoric water, and the variation in δ D value occurred due to the fractionation in hydrogen isotopes during boiling of geothermal waters [26].

6.4 Non-Traditional Isotopes in Geochemistry

The recent advancement in the techniques and the development of various advanced method have increased the applications of nontraditional isotopes in various fields. Significant information related to history and composition of geological or other samples can be derived by utilizing the metal isotopes. The most significant applications of metal isotopes are the tracing of processes and sources.

6.4.1 Application in Tracing of Source

Mixing of two different pools along with the isotopes of different compositions is the basic criteria of source tracing. The contribution of various sources in any sample can be calculated with the help of mixing calculations only if the isotopic compositions of the involved endmembers are significantly distinct and known. However, in isotope dilution technique, mixing of isotope can disturb the mass balance according to concentration data without affecting the isotope ratio. So, the calculation of delta value of a sample is given in Equation (6.7).

$$\delta_{\text{sample}} = \delta_{\text{pool A}} \times f_{\text{pool A}} + \delta_{\text{pool B}} \times f_{\text{pool B}} \tag{6.7}$$

where f represents the relative fraction of the reservoir or pool A and B.

The method of two pools mixing can illustrate the effect of isotope signatures and pool sizes on the isotopic composition of the mixture obtained. For instance, suppose, our soil sample has shown the isotopic composition of geogenic background of +0.5‰ and this soil sample was mixed with anthropogenic source sample which showed isotopic composition of −1.5‰; hence, the resulting isotopic composition of −1.0‰ would show 75% anthropogenic contribution. Another example which we can discuss here is the work of Chen *et al.* [27] where they examined the effect of anthropogenic (e.g., industrial or urban waste emissions) and natural (e.g., weathering of rocks) sources on river water (Seine River, France) by using the Zn isotopes. Till now, several metals have been used for source tracing purposes along with the signature of MIF also [28]. However, the researchers have also faced several difficulties during the source tracing in natural systems because of several reasons, i.e., (1) contribution of multiple sources into a sample, (2) lack of precise knowledge of isotopic compositions of mixing pools, and (3) effect of fractionating process on the isotopic composition of the sample. Therefore, for effective source tracing, the sample should have distinct and well-defined sources, and which should not be affected by fractionating process as well.

6.4.2 Application in Process Tracing

The concept of process tracing is that the target sample should pass through the transformation process which can alter the isotopic composition of sample. For instance, transformation process involved separation of solid from aqueous phase due to the transformation of soluble species into insoluble species such as in aqueous solution soluble Cr (VI) is reduced into the Cr (III) which is insoluble and hence, get precipitated in water [29]. The Rayleigh model has been very useful in this case, which can be used to quantify the extent of transformation reaction but the isotopic enrichment factor (ε) for the transformation reaction should be known. Equations of Rayleigh model can be represented in various forms, for example, the isotopic composition of reactant can be represented by Equation (6.8) of Rayleigh model as

$$\delta_{\text{reactant}} = \delta_0 + \varepsilon \ln f \tag{6.8}$$

where ε, δ_0, and f represent the enhancement factor and the fraction left, respectively. The isotopic composition of the instantaneous product can be determined by Equation (6.9):

$$\delta_{\text{inst. product}} = \delta_0 + \varepsilon \ln f + \varepsilon \tag{6.9}$$

Here, in process tracing, the dilution effect plays very important role in stable isotope method. For instance, decrease in the concentration of Cr (VI) in groundwater can be elucidated in two ways, i.e., the concentration of Cr (VI) decreases either due to reduction of Cr (VI) followed by its precipitation or by dilution of groundwater with unadulterated water. We cannot find out the extent of reduction only by the isotope ratios measurement [30]. Another example is the photoreduction Hg (II) in environmental samples where the isotopic signature of MIF and MDF both observed [31]. Process tracing and its quantitative application in environmental systems is challenging owing to the dependence of enrichment factor variability on transport processes and reaction conditions. Generally, under natural conditions, the enrichment factors show smaller value than the laboratory conditions [32]. Moreover, there are several other factors (e.g., mixing of pools of different isotopic compositions, simultaneous occurrence of processes with different isotope effects, and partial re-equilibration between reactant and product) which can make worse interpretation of isotopic compositions in natural samples [33]. In spite of several advancement have been made in early decades in process tracing, but the studies are still needed to find out the fractionation factors of unearthed. Consequently, the application of stable isotopes in complex system, process tracing, and quantification needed more experimental work for wide variety of elements.

6.4.3 Biological Cycling

In spite of organic elements, there are several metals which are essential for biological system and involved in biological cycling. From the beginning, the major motivation of non-traditional isotopic research was to find out the isotopic signature of biological activity in natural samples called the biosignature [34]. As we have discussed in earlier section (i.e., traditional isotopes in geochemistry), biologically controlled processes can also encourage substantial isotope fractionation in non-traditional isotopes similar to the traditional isotopes. Several biological processes involved enzymatic reaction for bond breaking (i.e., kinetically controlled) and thus inspire kinetic effects. Similar to the abiotic system, as discussed in above section, the kinetic effect in biological system also depends on pool sizes etc. Not only the kinetic effects, but the equilibrium effects also play crucial role in biological system. Several studies have been reported related to isotope fractionation of non-traditional isotopes during biologically mediated oxidation, assimilation, reduction, methylation, and demethylation [35–48]. Additionally, isotope fractionation in plants is also explored significantly, during uptake and translocation, and most of the studies have shown the enrichment of light isotopes in them with few exceptions [49–56]. Additionally, isotope fractionation in other higher organisms is also reported [57–59]. Consequently, the biological processes play very important role in environmental metal cycling and hence in assisted in metal isotope fractionation as well. Considering the earlier work related to traditional isotope fractionation will be helpful in further improvements in metal-based isotope fractionation studies [60].

6.5 Conclusion

Though the stable isotopes of traditional and non-traditional elements have shown versatile application in the field of geochemistry, yet, the importance of fractionation factor in metal isotope tracers is still lack with fruitful studies. However, since last decade, this scientific field has shown remarkable development. The different factors which control the isotope fractionation and various parameters have been explained for various non-traditional and traditional elements along with the combination of theoretical, experimental, and field-based studies which ultimately illustrate a comprehensive picture about the isotope geochemistry. But, in case of non-traditional isotopic studies, it is still new area of research in comparison to the traditional isotopic studies. So, the possibility of further development of new techniques and methods is still waiting in this field so that the several important scientific questions could be addressed. However, the use of metal isotope tracers in complex medium is still remain challenging due to availability of various metal sources in the sample and multiple processes involved in the analysis. Though, recent advancement in analytical techniques has increased the sensitivity of the current isotope methods and hence allowed measurement of natural sample at very low concentration and without pre-treatment as well.

References

1. Urey, H., The thermodynamic properties of isotopic substances. *J. Chem. Soc. (London)*, 1947, 562–581, 1947.
2. Bigeleisen, J. and Mayer, M.G., Calculation of equilibrium constants for isotopic exchange reactions. *J. Chem. Phys.*, 15, 261–267, 1947.
3. Hoefs, J., *Stable Isotope Geochemistry*, Springer, Berlin, 2009.
4. Fry, B., *Stable Isotope Ecology*, Springer, New York, 2006.
5. *Stable Isotopes in Ecology and Environmental Science*, Michener, R. and Lajtha, K. (Eds.), Blackwell, Oxford, 2007.
6. *Handbook of Stable Isotope Analytical Techniques*, de Groot, P.A. (Ed.), Elsevier, Amsterdam, 2008.
7. Johnson, C.M., Beard, B.L., Albarede, F., *Geochemistry of Non-Traditional Stable Isotopes: Reviews in Mineralogy Volume*, p. 454, Mineralogical Society of America, Washington, 2004.
8. Coplen, T.B., Guidelines and recommended terms for expression of stable-isotope-ratio and gas-ratio measurement results. *Rapid Commun. Mass Spectrom.*, 25, 2538–2560, 2011.

9. Criss, B.E., *Principles of Stable Isotope Distribution*, Oxford University Press, New York, 1999.
10. Schauble, E., Applying stable isotope fractionation theory to new systems. *Rev. Mineral. Geochem.*, 55, 65–112, 2004.
11. Schauble, E.A., Méheut, M., Hill, P.S., Combining metal stable isotope fractionation theory with experiments. *Elements*, 5, 369–374, 2009.
12. Wiederhold, J.G., Cramer, C.J., Daniel, K., Infante, I., Bourdon, B., Kretzschmar, R., Equilibrium mercury isotope fractionation between dissolved Hg(II) species and thiol-bound Hg. *Environ. Sci. Technol.*, 44, 4191–4197, 2010.
13. W. M. White. Stable isotope geochemistry. *Geochemistry*, USA, 701, 2005.
14. Farquhar, J. and Wing, B.A., Multiple sulfur isotopes and the evolution of atmospheric oxygen. *Earth Planet. Sci. Lett.*, 213, 1–13, 2003.
15. Johnson, D.G., Jucks, K.W., Traub, W.A., Chance, K.V., Isotopic composition of stratospheric ozone. *J. Geophys. Res.*, 105, 9025–9031, 2001.
16. Chiba, H., Chacko, T., Clayton, R.N., Goldsmith, J.R., Oxygen isotope fractionations involving diopside, forsterite, magnetite, and calcite: Application to geothermometry. *Geochim. Cosmochim. Acta*, 53, 2985–2995, 1989.
17. Park, R. and Epstein, S., Carbon isotope fractionation during photosynthesis. *Geochim. Cosmochim. Acta*, 21, 110–126, 1960.
18. O'Leary, M.H., Carbon isotope fractionation in plants. *Phytochemistry*, 20, 553–567, 1981.
19. DeNiro, M.J. and Hasdorf, C.A., Alteration of 15N/14N and 13C/12C ratios of plant matter during the initial stages of diagenesis: studies utilizing archeological specimens from Peru. *Geochim. Cosmochim. Acta*, 49, 97–115, 1985.
20. DeNiro, M.J. and Epstein, S., Influence of diet on the distribution of nitrogen isotopes in animals. *Geochim. Cosmochim. Acta*, 45, 341–351, 1981.
21. DeNiro, M.J. and Epstein, S., Influence of diet on the distribution of carbon isotopes in animals. *Geochim. Cosmochim. Acta*, 42, 495–506, 1978.
22. Schidlowski, M., A 3800-million-year isotopic record of life from carbon in sedimentary rocks. *Nature*, 333, 313-318, 1988.
23. Mojzsis, S.J., Arrhenius, G., McKeegan, K.D., Harrison, T.M., Nutman, A.P., Friend, C.R.L., Evidence for life on Earth before 3800 million years ago. *Nature*, 384, 55–59, 1996.
24. Rosing, M., 13C-depleted carbon microparticles in >3700 Ma sea-floor sedimentary rocks from West Greenland. *Science*, 283, 674–676, 1999.
25. Craig, H., The isotopic composition of water and carbon in geothermal areas, in: *Nuclear Geology on Geothermal Areas*, E. Tongiorgi (Ed.), pp. 17–53, CNR Lab. Geol. Nucl., Pisa, 1963.
26. Ohmoto, H., Stable isotope geochemistry of ore deposits, in Stable Isotopes in High Temperature Geological Processes, in: *Reviews in Mineralogy*, vol. 16, J.W. Valley, H.P. Taylor, J.R. O'Neil (Eds.), pp. 491–560, Mineral. Soc. Am., Washington, 1986.

27. Chen, J.B., Gaillardet, J., Louvat, P., Zinc isotopes in the Seine River waters, France: A probe of anthropogenic contamination. *Environ. Sci. Technol.*, 42, 6494–6501, 2008.
28. Estrade, N., Carignan, J., Donard, O.F.X., Isotope tracing of atmospheric mercury sources in an urban area of northeastern France. *Environ. Sci. Technol.*, 44, 6062–6067, 2010.
29. Ellis, A.S., Johnson, T.M., Bullen, T.D., Chromium isotopes and the fate of hexavalent chromium in the environment. *Science*, 295, 2060–2062, 2002.
30. Berna, E.C., Johnson, T.M., Makdisi, R.S., Basu, A., Cr stable isotopes as indicators of Cr(VI) reduction in groundwater: A detailed time-series study of a point-source plume. *Environ. Sci. Technol.*, 44, 1043–1048, 2010.
31. Sherman, L.S., Blum, J.D., Johnson, K.P., Keeler, G.J., Barres, J.A., Douglas, T.A., Mass-independent fractionation of mercury isotopes in arctic snow driven by sunlight. *Nat. Geosci.*, 3, 173–177, 2010.
32. Clark, S.K. and Johnson, T.M., Effective isotopic fractionation factors for solute removal by reactive sediments: A laboratory microcosm and slurry study. *Environ. Sci. Technol.*, 42, 7850–7855, 2008.
33. Lutz, S.R. and Van Breukelen, B.M., Combined source apportionment and degradation quantification of organic pollutants with CSIA: 1. Model derivation. *Environ. Sci. Technol.*, 48, 6220–6228, 2014.
34. Beard, B.L., Johnson, C.M., Cox, L., Sun, H., Nealson, K.H., Aguilar, C., Iron isotope biosignatures. *Science*, 285, 1889–1892, 1999.
35. Beard, B.L., Johnson, C.M., Cox, L., Sun, H., Nealson, K.H., Aguilar, C., Iron isotope biosignatures. *Science*, 285, 1889–1892, 1999.
36. Ader, M., Chaudhuri, S., Coates, J.D., Coleman, M., Microbial perchlorate reduction: A precise laboratory determination of the chlorine isotope fractionation and its possible biochemical basis. *Earth Planet. Sci. Lett.*, 269, 604–612, 2008.
37. Crosby, H.A., Roden, E.E., Johnson, C.M., Beard, B.L., The mechanisms of iron isotope fractionation produced during dissimilatory Fe(III) reduction by Shewanella putrefaciens and Geobacter sulfurreducens. *Geobiology*, 5, 169–189, 2007.
38. Sikora, E.R., Johnson, T.M., Bullen, T.D., Microbial mass dependent fractionation of chromium isotopes. *Geochim. Cosmochim. Acta*, 72, 3631–3641, 2008.
39. Croal, L.R., Johnson, C.M., Beard, B.L., Newman, D.K., Iron isotope fractionation by Fe(II)-oxidizing photoautotrophic bacteria. *Geochim. Cosmochim. Acta*, 68, 1227–1242, 2004.
40. Kappler, A., Johnson, C.M., Crosby, H.A., Beard, B.L., Newman, D.K., Evidence for equilibrium iron isotope fractionation by nitrate-reducing iron(II)-oxidizing bacteria. *Geochim. Cosmochim. Acta*, 74, 2826–2842, 2010.
41. Schilling, K., Johnson, T.M., Wilcke, W., Isotope fractionation of selenium during fungal biomethylation by Alternaria alternata. *Environ. Sci. Technol.*, 45, 2670–2676, 2011.

42. Rodriguez-Gonzalez, P., Epov, V.N., Bridou, R., Tessier, E., Guyoneaud, R., Monperrus, M., Amouroux, D., Species-specfic stable isotope fractionation of mercury during Hg(II) methylation by an anaerobic bacteria (Desulfobulbus propionicus) under dark conditions. *Environ. Sci. Technol.*, 43, 9183–9188, 2009.
43. Perrot, V., Bridou, R., Pedrero, Z., Guyoneaud, R., Monperrus, M., Amouroux, D., Identical Hg isotope mass dependent fractionation signature during methylation by sulfate-reducing bacteria in sulfate and sulfate-free environment. *Environ. Sci. Technol.*, 49, 1365–1373, 2015.
44. Kritee, K., Blum, J.D., Reinfelder, J.R., Barkay, T., Microbial stable isotope fractionation of mercury: A synthesis of present understanding and future directions. *Chem. Geol.*, 336, 13–25, 2013.
45. Wasylenki, L.E., Anbar, A.D., Liermann, L.J., Mathur, R., Gordon, G.W., Brantley, S.L., Isotope fractionation during microbial metal uptake measured by MC-ICP-MS. *J. Anal. At. Spectrom.*, 22, 905–910, 2007.
46. Cameron, V., Vance, D., Archer, C., House, C.H., A biomarker based on the stable isotopes of nickel. *Proc. Natl. Acad. Sci. U.S.A.*, 106, 10944–10948, 2009.
47. John, S.G., Geis, R.W., Saito, M.A., Boyle, E.A., Zinc isotope fractionation during high-affinity and low-affinity zinc transport by the marine diatom Thalassiosira oceanica. *Limnol. Oceanogr.*, 52, 2710–2714, 2007.
48. Horner, T.J., Lee, R.B.Y., Henderson, G.M., Rickaby, R.E.M., Nonspecific uptake and homeostasis drive the oceanic cadmium cycle. *Proc. Natl. Acad. Sci. U.S.A.*, 110, 2500–2505, 2013.
49. Kiczka, M., Wiederhold, J.G., Kraemer, S.M., Bourdon, B., Kretzschmar, R., Iron isotope fractionation during Fe uptake and translocation in alpine plants. *Environ. Sci. Technol.*, 44, 6144–6150, 2010.
50. Deng, T-H-B., Cloquet, C., Tang, Y.-T., Sterckeman, T., Echevarria, G., Estrade, N., Morel, J.-L., Qiu, R.L., Nickel and zinc isotope fractionation in hyperaccumulating and non-accumulating plants. *Environ. Sci. Technol.*, 48, 11926–11933, 2014.
51. Jouvin, D., Weiss, D.J., Mason, T.F.M., Bravin, M., Louvat, P., Zhao, F.J., Ferec, F., Hinsinger, P., Benedetti, M.F., Stable isotopes of Cu and Zn in higher plants: Evidence for Cu reduction at the root surface and two conceptual models for isotope fractionation processes. *Environ. Sci. Technol.*, 46, 2652–2660, 2012.
52. Ryan, B.M., Kirby, J.K., Degryse, F., Harris, H., McLaughlin, M.J., Scheiderich, K., Copper speciation and isotopic fractionation in plants: Uptake and translocation mechanisms. *New Phytol.*, 199, 367–378, 2013.
53. Weiss, D.J., Mason, T.F.D., Zhao, F.J., Kirk, G.J.D., Coles, B.J., Horstwood, M.S.A., Isotopic discrimination of zinc in higher plants. *New Phytol.*, 165, 703–710, 2005.

54. Smolders, E., Versieren, L., Dong, S., Mattielli, N., Weiss, D.J., Petrov, I., Degryse, F., Isotopic fractionation of Zn in tomato plants reveals the role of root exudates on Zn uptake. *Plant Soil*, 370, 605–613, 2013.
55. Martin, J.E., Vance, D., Balter, V., Magnesium stable isotope ecology using mammal tooth enamel. *Proc. Natl. Acad. Sci. U.S.A.*, 112, 430–435, 2015.
56. Walczyk, T. and von Blanckenburg, F., Natural iron isotope variations in human blood. *Science*, 295, 2065–2067, 2002.
57. Walczyk, T. and von Blanckenburg, F., Deciphering the iron isotope message of the human body. *Int. J. Mass Spectrom.*, 242, 117–134, 2005.
58. von Blanckenburg, F., Noordmann, J., Guelke-Stelling, M., The iron stable isotope fingerprint of the human diet. *J. Agric. Food Chem.*, 61, 11893–11899, 2013.
59. Jaouen, K., Pons, M.-L., Balter, V., Iron, copper and zinc isotopic fractionation up mammal trophic chains. *Earth Planet. Sci. Lett.*, 374, 164–172, 2013.
60. Hayes, J.M., Fractionation of carbon and hydrogen isotopes in biosynthetic processes. *Rev. Mineral. Geochem.*, 43, 225–277, 2001.

7

Environmental Geochemistry

Sapna Nehra[1], Rekha Sharma[2] and Dinesh Kumar[3]*

[1]*Department of Chemistry, Dr. K.N. Modi University, Newai, Rajasthan, India*
[2]*Department of Chemistry, Banasthali Vidyapith, Banasthali, Rajasthan, India*
[3]*School of Chemical Sciences, Central University of Gujarat, Gandhinagar, India*

Abstract
To investigate the pollution caused in the environment, various geochemical factors are responsible, which comprises a higher concentration of heavy metals, weathering of rocks, mine tailings, and so forth. In geochemistry, investigation finds out the main factor which causes pollution to the environment in particular regions. Many researchers choose the basin, mining area, river, and mountains to check its geochemical behavior regarding the pH, TDS, EC, heavy metals content, minerals, clay, and rocks constituents. With the help of analytical tools, the physicochemical properties of particular geochemical features are characterized. This chapter presented the overview of last decades' research outcomes to know how geochemistry affects the environment stability and causes pollution at various levels.

***Keywords*:** Mining, geochemistry, weathered tailing, pH, total dissolved solids, electrical conductivity

7.1 Introduction

Worldwide, the mining and ore extraction generated a huge number of toxic byproducts, which heavily affects the environment and human health. Because of physical, geochemical, and mineralogical features, the tailing mine behavior as the capable source of nearby contamination like airborne transmittance of the tiny particles at the time of dryness is confirmed [1, 2]. The machine-driven transference of tailings through

**Corresponding author*: dinesh.kumar@cug.ac.in

Inamuddin, Mohd Imran Ahamed, Rajender Boddula and Tariq Altalhi (eds.) Geochemistry: Concepts and Applications, (111–126) © 2021 Scrivener Publishing LLC

the water flows throughout modest to extreme rain falls. Further transfer of the soluble hazardous substances in acidic leachates. Basic factors which activate the danger of ecological contamination are shown because of oxidation of S^{2-} specifically in pyrite form [3–7]. The mine products are protecting through the water at the time of mining operations. In terms of long-lasting, the water layers can vanish, which permits the admittance of environmentally O_2 into S^{2-}. As a result, the well-gained sulfide passes through the oxidation, which abruptly in the group of more acidic water with a maximum quantity of sulfate and significant hazardous elements called AMD [8–10]. The yellow, reddish appearance of the ting pond occurred due to existence of iron (III), oxyhydroxides, crusts, and hydrated sulfates, hardpans [11–14]. Besides the major elements, it comprises the iron in both state +3 and +2 state aluminum, magnesium, and hydrated sulfates that behave as the trapper for the potentially toxic elements released via the AMD, for example, copper, lead, zinc, cadmium, and arsenic [15–40].

7.2 Overview of the Environmental Geochemistry

Morales-Simfors *et al.* reviewed the geochemistry of the arsenic origin and its sources like surface and groundwater and soil, which negatively impacts the environment. Several natural processes are responsible for the production of arsenic. In Latin America, a different geothermal and the volcanic process contributed to the origination of the arsenic, which adversely affects the human population. This review summarized the 15 country data of Latin America. They have investigated the 423 sites with the presence of arsenic, which mainly originated from the geothermal sources. Mostly, volcanic activity is responsible for its origin. Several geothermal systems and the volcanoes counted in high populated cities comprise San José, Guatemala, Bogota, Managua, and Mexico City. In these cities, the concentration of the arsenic exists beyond the recommended limit of 0.01 mg/L by WHO. Although, only based on geographical structure cannot find out the origin of arsenic in geothermal fluids and volcanic emissions. Therefore, it very urgent to develop geographical maps based on GIS to find out the real cause of the occurrence of arsenic. The studies related to environmental threats and adverse impacts on human health are very rare. So, it should notify this present research gap in future ongoing research. Existing all-inclusive assessment of arsenic creating from geothermal characteristics and volcanic emissions along with their movement, geochemistry, and availability can be powerful to design a plan to found a justifiable arsenic removal in susceptible areas of Latin America in the near future [41].

Zouhri *et al.* studied the wastewater of Kettara village, with uncontrolled investigations, establish a possible ecological problem that directly affects the quality of water, air, soil, and human health. Here, the performed test concluded the ecological impression brought by mine stakeouts, lithological factors of rocks, and wastewater. So, present multiple field research permits knowing the transfer of the metallic trace elements like As, Ni, Zn, Se, and SO_4^{2-} ions in the cracked shales medium. To measure the water safety level by checking through the microbiological investigation. Subsequently, this analysis predicts because of the availability of different bacteria like Fecal coliforms, E. coli, Enterococci, coliforms, Salmonella, Mesophilic, and Aerobic Flora, Sulfite-reducing bacteria in groundwater resources influences at the domestic level. Se, As, and the pathological pollution can be described by five features.

(1) It depends on the geological constructions and the nature of the hydrogeological scheme.
(2) No more availability of a healthy environment because there is no suitable old-style septic reservoir situated in the village.
(3) The passing of animal excreta easily reaches to groundwater from the soil.
(4) The movement of the wastewater toward groundwater generates pollution. The presence of the reed beds can be explained the declination of the bacteria in the hydrogeological wells of the examined area [42].

Battioui *et al.* reported the after producing 20 million tonnes of solid waste by Jerada anthracite mine in Morocco in 2001 were unrestricted. By testing the different collecting samples from the waste accumulation and tested the acid-forming tendency. During acid testing, no carbonate formation was observed. Only the sulfide mineral pyrite was identified, which largely converted into the iron oxides. The testing results of waste show the minimum level of the calcium and magnesium of 1%, whereas iron and aluminum content usually over 5%. Altered Sobek static and column assessment was performed for the period of 24 months. The kinetics results of the tested three samples showed initially acidic pH the out of five samples. Later on, at the ending of the test, remaining two samples also showed acidic behavior. With the time the concentration of the SO_4^{2-} reduced from 5,000 to 200 mg/L [43].

Wilson *et al.* showed the Diavik Waste Rock Project, which includes both the laboratory and the field study for the evaluation of the geochemical progression of S^{2-} mine wastes. In this study, the humidity cell tests to describe the long-lasting development of low S^{2-} waste rock are firstly performed. The reactive transport modeling was utilized to measure the substantial geochemical evolution that determines the oxidation of S^{2-} minerals, and it required both temperature and quantity of mineral. The content of different metals and ions was found as nickel, cobalt, copper, zinc, and sulfate ion in moisture seepage cells that were replicated by using the lessening essential prototypical, which signified the regulator of oxidant diffusion to the unreactive surface element in the S^{2-} oxidation method. The effect of temperature was determined by using the Arrhenius relation and suitable activation energy data. By analyzing the experimental outcomes, which contains the waste rock discriminated by S^{2-} quantity and temperature, assigned surface area, and main temperature role in the rate of S^{2-} oxidation and elimination of SO_4^{2-} and metals. Afterward, the calibration of the model found it better fit with the seepage value gained from the S^{2-} content cells. Subsequently, replications were performed via managing the estimated factors comprises S^{2-} concentration and surface area [44].

Stumbea *et al.* closely observed a pond that accumulated by the byproducts got from the mining at the northern part of the beach. Here, mainly three different waste byproducts were assessed. First rename as A-type shows strong oxidation and evaporation where pH lies between 2.2 and 2.6, and SF was 14.2%. The second category as B-type contains byproduct from lowland waste and pH value varied from 2.7 to 3.1, and SF value 7.7%. The third category salt crusts denoted as the C-type generated by the vanishing of leachates gathered in stationary pools where pH ranges from the 2.6 to 3.5, and SF value was 81.5%. The PCA analysis displayed as subsequent orders of subordinate substances coagulation.

The A type consists of the ferricopiapite, rozenite, ferrihydrite, jarosite, coquimbite; B type is made up of iron oxyhydroxides, butlerite, jarosite; and C type consists of epsomite, hexahydrite, rozenite ferrihydrite, and alunogen.

This analysis showed the aluminum from 1% to 10% and 1% to 5 % in the form of aluminum oxide in A, B, and C type waste. The iron content lies between 11% and 23% in the form of iron oxide 5% to 10% in A, B, and C waste. Zinc and cadmium were obtained from the ferricopiapite, rozenite, jarosite, halotrichite, and apjohnite, although copper, lead, and arsenic were captured by the clay minerals [45].

Shahhosseini *et al.* examined the geochemistry of the arsenic and the lead in the dumped product of the copper mine situated in Iran. With the help of different characterization like XRD, SEM-EDS, EPMA, RS, assessed the pyrite waste. It estimated the spreading of arsenic and the lead to the degree of their mobilization. The core amount of the arsenic and lead in mean, minimum, and maximum order was 27.28, 12.4, 59.5, and 72.05, 19, 159 mg/kg, respectively. These minerals are linked with the pyrite crystal and the bounded form of galena in pyrite. Because of the presence of the higher content of the sulfate ion in range of 3,950 to 286,000 mg/L, established the evidenced oxidation of the pyrite. In the waste byproduct leached constituents, the pH value lies in 1.47 to 4.23 due to the lessening of carbonate minerals in the dumped waste. The accumulation of secondary minerals gained through the weathering, together with plumbojarosite, brochantite, and coquimbite, was recognized. The linkage arsenic and lead with hydroxosulfates were identified. Therefore, the existence of lead and arsenic jarosite minerals was the good regulatory factor with a minimum quantity of arsenic and lead lies from 0.05 to 8.25 and 0.07 to 1.61 mg/L, respectively. The maximum dissolution analyzed for arsenic and lead at pH 2 [46].

Zhang *et al.* illustrated the environmental geochemistry features of groundwater in Huainan Coalfield, southern China. They collected the 59 test sample from 12 sites and about 18 heavy metals like lithium, vanadium, copper, lead, zinc, cadmium, nickel, manganese, cobalt, chromium, iron, barium, titanium, strontium, molybdenum, and arsenic. The seven other ions that included potassium, sodium, calcium, sodium, magnesium, chloride, sulfate, and bicarbonate are also estimated. According to the People's Republic of China Quality Standard, the concentration of the iron and manganese lies in the class III. The levels of lead, zinc, cadmium, chromium, cobalt, and barium range in class II are estimated by the fuzzy mathematical process. With the 11 heavy metals positive pH correlated as comprises lithium, vanadium, copper, lead, zinc, cadmium, nickel, manganese, chromium, barium, strontium, and aluminum, titanium, and cobalt, respectively. The order of loading rates influenced by the pH sequentially in increasing order was aluminum > barium > titanium > silicon > lithium > manganese. The lithium quantity was linked with six ions such as sodium, calcium, magnesium, chloride, sulfate, and bicarbonate, and strontium was associated with the first five ions, and Barium was linked with first four ions. PCA analysis of vanadium, arsenic, manganese, lead, cadmium, nickel, chromium, lithium, and zinc in the groundwater revealed that it could be elucidated from leaching of soil, rocks, surface runoff, precipitation and surface pollution by heavy metals [47].

Sharafi *et al.* explained the geochemistry of rare earth metals and major and traced metals nearby region of the Angouran zinc-lead mine. To simulate the data of the thermodynamics, PHREEQC module was utilized. Tailing water samples of the Calcimin zinc lead processing plant which in Zanjan province shows the pH level from 8 to 7.9, and also exhibited that the content of sulfate, zinc, lead, arsenic, cadmium was found to be 1,179–15,000 mg/L, 18,529 mg/L, 1780 μg/L, 465 μg/L, and 230 μg/L, respectively. However, the concentration varied in Angouran pit samples like pH value presented between 7.6 and 8.25 means shows neutral and alkaline medium, and smaller content of sulfate was 21 to 87 mg/L and raised concentration zinc till 27 mg/L, lead and cadmium till 586.3, and 33 μg/L, correspondingly. The pH of water expels from the Angouran mine adit was 7.5 to 8.05 with alteration of sulfate, zinc, lead, and arsenic concentration from 30 to 2294 mg/L, 8 mg/L, 123 μg/L, and 35.4 μg/L, respectively. According to the USEPA, the concentration of heavy metals in plant waters includes arsenic, lead, cadmium, nickel, and zinc exceeds their limit. Usually, the variety of water tailing from plant and mine ranged from $ZnSO_4$, $NaKSO_4$ to $CaHCO_3$, correspondingly. Mostly, the hazardous metals comprise lead, cadmium, nickel, and zinc indicated the cationic form in the plant water [48].

Khorasanipour *et al.* presented the geoenvironmental nature of the REEs in the Sarcheshmeh mine tailing pond, located in the semiarid environment. To check the geochemistry characteristic, solid and water samples were taken which possess:

(i) weathering and oxidation processes and ascendant movement of soluble constituents;
(ii) equilibrium and creation and termination of the secondary evaporative phases;
(iii) in the surface and groundwater resources, many hydrogeochemical alterations;
(iv) flotation of the copper S^{2-} ore and tailing regulatory approaches.

The average REEs were 113.0, 103.7, and 74.6 mg/kg for the fresh, weathered tailings, and the evaporative medium. On the comparison of fresh, weathered, and the evaporative phases, the substantial augmentation of heavy rare earth metals is displayed and well depleted pattern of light rare earth metals except for europium. The geochemical feature predicts the higher dissolvability and agility tendency of heavy rare earth metals and europium equated with light rare earth metals, which results in the better quantity and fulfillment of significant elements of the evaporative surface coating created on surface weathered tailings. The obtained water surface comprises the long range of hydrogeochemical characteristics like acidic, neutral, and alkaline pH from 2.5–4.7, 6.4–7.9, and 8–10, and highly alkaline > pH 10. In acidic water, the rare earth element concentration was found to be 489 to 1903 μg/L, which were generated from weathered tailings material resulted from heavy rainfall. However, the average of rare earth elements reduced with the raising pH. This follow up remains the same with the maximum content of heavy rare earth elements relatively than light rare earth elements. The speciation values that revealed the $LnSO_4^+$, $Ln(SO_4)^{2-}$, and Ln^{3+} are the principal forms of soluble rare earth elements in acidic and around neutral pH, and the dominant species were found in different percentage in the presence of alkaline medium 40.78%, 31.68%, and 31.07% as $LnCO_3^+$, $Ln(CO_3)^{2-}$, and $LnSO_4^+$, respectively. But, in more alkaline pH condition speciation design altered to $LnOH^{2+}$, $Ln(CO_3)^{2-}$, and $LnCO_3^+$ as 53.15%, 44.88%, 2.61%. Here, Ln denotes the lanthanide. Based on the previous investigations concluded, the ionic potential attributed the chief role in the determination of the rare earth elements adsorption potential, agility, and ionic replacement [49].

Van der Ent *et al.* investigated the MCM in Malaysia on Borneo Island. Here, the mining process was operated from the year 1975 to 1999. During the mining process, overall, 2.47 metric ton solid make up was produced, which further comprise the different elements like the copper, gold, and silver as 45, 294, and 294 ton, respectively. It produced the 250 metric ton residue with the rock waste over 150 metric ton tailing, which accumulated at the 397 ha Lohan tailings storage capability, 15.8 km from the mine, and 980 m below in altitude. MCM site attributed the task for the ecological therapy because of the presence of the maximum sulfide minerals wastes, heavy rainfall, and more volume of contaminated mine pit water. It revealed the therapy and treatment consumes high cost as to neutralize the acidic mine pit effluent required more amount of the lime. At the MCM site, various infrequent geochemical characteristics on the reason of the associated existence of acid-producing S^{2-} rocks and limy minerals like serpentinite and exhibited exceptional biotic characteristics due presence of vast flora variability. This unique site offers a valued chance for investigating ordinary acid neutralization procedures and mine restoration in tropical areas.

These days, this site is shielded by nature reserves like Kinabalu Park and Bukit Hampuan. The natural ecological heritage inhibited the degazetting and inclusion in this endangered area in the conceivable future [50].

Tlil *et al.* illustrated the geochemistry of lead-zinc mining, which contains the number of toxic elements in the district of Northern Tunisia. The toxic elements were an account of zinc, lead, cadmium, arsenic, and antimony as 6.9 to 13.7, 2.9 to 9.6, 31 to 106, and 180 to 837 mg/kg, respectively. Mineralogical investigation showed the primary S^{2-}, sulfosalts, i.e., sphalerite, chalcopyrite, arsenopyrite galena, pyrite, and sulfosalts, and secondary oxidized constituents in the form of minerals, and sulfates like smithsonite cerussite, and barite gypsum, correspondingly. With all these clay matrices, calcite was found. Some experiments like the leaching assessment with distilled water showed that the reaction occurred at a neutral point with the liberation of potentially toxic elements such as arsenic, antimony, zinc, cadmium, and sulfate between 16 and 58.5, 1.5 and 60, 222 and 1,130, 4,865 and 15,665, 4 and 25, and 3690 and 22,070 μg/L, respectively. All the above-mentioned concentration was high than WHO limits. So, this kind of polluted water causes many threats to the environment. Therefore, it should be properly treated and examined [51].

Essilfie-Dughan *et al.* showed the acquaintance of coal waste rock with the surrounding atmosphere, which results in the liberation of the SO_4^{2-} and some other elements like arsenic, cadmium, and zinc in the ground and the surface water through an oxidation process. Likewise, the Fe-containing minerals like ankerite, sulfide, siderite, and pyrite are diminished. These constituents are present in the waste rock, which can easily form the oxyhydroxides and which adsorb trace elements released by the oxidation of S^{2-}. The analytical investigations performed to measure the quantity of the dispersed minerals of S^{2-}, Fe, and its additional oxidation product, leaching rate. All these investigations have concluded the effect of sulfate and traces elements to nearby ground and surface water. Waste rock at Elk Valley, British Columbia was analyzed by the synchrotron-based X-ray absorption spectroscopy and found that the average iron content as pyrite, siderite, goethite, and lepidocrocite was 6.0%, 44.3%, 35.4%, and 14.3% with S^{2-} average ratio, organic S^{2-}, 26.9, and sulfate was 26.9, 58.7, and 14.4, respectively. In the squeezed porewater test, obtained concentration of iron and sulfate was 0.28 and 1246 mg/L, respectively. The porewater collected from the dump were supersaturated by the iron oxyhydroxides and undersaturated by gypsum. The coupling of iron and sulfur mineralogical information with long-lasting water excellence and quantity dimensions that indicated the 10% of the S^{2-} which comprise the 2% of the sulfur in the landfill were oxidized in over 30 years of a time. The sulfur obtained after oxidation of S^{2-} was liberated to getting surface water as sulfate and iron precipitated as iron oxyhydroxides by elimination of iron around 3.0×10^{-5}% waters over the past 30 years. However, the results show the leakage of sulfate, considered no alteration in the speed of oxidation of S^{2-}, and sulfate was not a major issue in collected surface waste samples because it presents at below level according to recommended limits [52].

Li *et al.* demonstrated the geochemical response of antimony at the Duliujiang river basin, positioned in southwest China, and check its environmental pollution impact. Antimony leans toward to migrate in the dissolvable state under the water-sediment system, but the fortune of soluble antimony was not well understood. This basin contributed to the natural environment to examine the passage of antimony due to their distinct geographical and geological features. Various other physicochemical characteristics include the pH, electrical potential, electrical conductivity, dissolved oxygen, flux, trace elements as antimony, arsenic, strontium, and main ions such as calcium, magnesium, and sulfate. These elements' quantity influences the water quality and regulates the dispersal and transference capability. During the experiment, collected three different varieties of waters like mainstream waters, tributary waters, and adit waters show the pH as 7.33 to 8.43, 6.85 to 9.12, and 7.57 to 9.76, correspondingly. Among all three verities higher quantity of antimony found in the adit water, which exhibited the 13,350 μg/L released from the antimony mine, and the mine waste shows the 8,792 mg/kg antimony obtained from the historical mine landfills. Soluble antimony presented the sturdy migration tendency in streams, meanwhile mainly relies on the less concentrated feature of stream water exhibited maximum flow speed. The exodus region of the basin shows 10 μg/L of mean antimony content. It formed a very smooth DI water extraction method to determine the tendency of the antimony solubility from the mine wastes. The obtained outcomes illustrated that the maximum solubility was found in an alkaline medium in which pH range lies from 7.11 to 8.16 compared to acidic pH that was 3.03–4.45 of mine wastes. It indicated that the mine wastes that confined more antimony contents could be liberated in the ordinary stream water system [53].

Aravinthasamy *et al.* estimated the geochemical behavior as well as health impact of F^- pollution on that person who lived at the Shanmuganadhi River basin in Tamil Nadu. From the wells which used for drinking purpose around 61 water samples were taken at the time of post-and pre-monsoon seasons. Subsequently, it undergoes the analysis to verify the wide physicochemical constraints. Mainly the characteristics were illustrated like pH, TDS, EC, and also the presence of the other ions as calcium, magnesium, sodium, potassium, bicarbonate, chloride, sulfate, nitrate, phosphate, and fluoride. In the monsoon season, concentration of F^- in tested samples of examined sites diverse from 0.01 to 2.50 and 0.01 to 3.30 mg/L, correspondingly. Among 61 water samples fixed fourteen in the post-monsoon period, and 16 samples in pre-monsoon show the high concentration of F^- attributed the dental fluorosis in a particular region. In that region, the granitic and gneissic rocks that include hornblende, apatite, biotite muscovite, and amphiboles were mainly responsible for the elimination of fluoride. Besides these natural sources, anthropogenic activities also attributed to access the fluoride to groundwater by using the fertilizer and pesticides. The network of dispersion of the fluoride in diverse natural water resources revealed that the well in charnockite terrain exhibited the very less F^- content than the well situated in the hornblende–biotite gneiss. So, possibilities of causing dental and skeletal fluorosis in those regions largely increased. As well F^- persists positive association with HCO_3^- in groundwater. So, prevalent fluoride areas usually gain a large amount of HCO_3^- in groundwater samples of Shanmuganadhi basin comes in the endemic category. So, this study investigated the in Shanmuganadhi basin around 26 villages at a higher risk of fluorosis, which required attention to treat the fluorinated water and make them suitable for drinking [54].

7.3 Conclusions

Despite the several investigations at different sites located in the basin, rocks, rivers, and in plains, it should employ the efficient method to save the environment from the pollution. The discussed reports confirmed the heavy metal reaches to groundwater and surface water from through the leaching of rock minerals, volcanic rocks, and mine tailings. But, behind these all processes, the physicochemical mechanism is still unknown. The effect of dry and wet seasons on transferring the heavy metals and other major constituents. This means that the most serious period of the season should be determined. Also, the different modes of the mobility of toxic pollutant in the local communities where the people live at their high risk are investigated. At the higher risk of exposure with metalloids in which arsenic is one of them for that specifically, the zero-valent iron combined with the citrate and solar light is used. Utilization of organic could inhibit the higher concentration of metals like copper, which transfer into the food chain by the irrigation process in crop fields.

7.4 Abbreviations

AMD	Acid mine drainage
WHO	World health organization
GIS	Geographic information system
S^{2-}	sulfide
SF	Soluble fraction average
PCA	Principal component analysis
XRD	X-ray diffraction
SEM	Scanning electron microscope
EPMA	Electron probe micro-analysis
Raman Spectroscopy	RS
NASC	North-American Shale Composite
REEs	Rare earth elements
LREEs	Light rare earth elements
HREEs	Heavy rare earth elements
Mt	Metric ton
MCM	Mamut copper mine
TDS	Total dissolved solids
EC	Electrical conductivity
BCR	Community Bureau of Reference, European Commission
WAS	World average shale
AC	Average chondrite

Acknowledgment

Dinesh Kumar is thankful to DST, New Delhi, for financial support to this work (sanctioned vide project Sanction Order F. No. DST/TM/WTI/WIC/2K17/124(C).

References

1. Ojelede, M.E., Annegarn, H.J., Knee, M.A., Evaluation of aeolian emissions from gold mine tailings on the Witwatersrand. *Aeolian Res.*, 3, 477, 2012.
2. Corriveau, M.C., Jamieson, H.E., Parsons, M.B., Campbell, J.L., Lanzirotti, A., Direct characterization of airborne particles associated with arsenic-rich mine tailings: Particle size, mineralogy and texture. *Appl. Geochem.*, 26, 1639, 2011.

3. Rico, M., Benito, G., Salgueiro, A.R., Díez-Herrero, A., Pereira, H.G., Reported tailings dam failures. A review of the European incidents in the worldwide context. *J. Hazard. Mater.*, 152, 846, 2008.
4. Moncur, M.C., Ptacek, C.J., Blowes, D.W., Jambor, J.L., Release, transport and attenuation of metals from an old tailings impoundment. *Appl. Geochem.*, 20, 639, 2005.
5. Romero, F.M., Armienta, M.A., González-Hernández, G., Solid-phase control on the mobility of potentially toxic elements in an abandoned lead/zinc mine tailings impoundment Taxco Mexico. *Appl. Geochem.*, 22, 109, 2007.
6. Bea, S.A., Ayora, C., Carrera, J., Saaltink, M.W., Dold, B., Geochemical and environmental controls on the genesis of soluble efflorescent salts in coastal mine tailings deposits: A discussion based on reactive transport modeling. *J. Contam. Hydrol.*, 111, 65, 2010.
7. Quispe, D., Pérez-López, R., Acero, P., Ayora, C., Nieto, J.M., The role of mineralogy on element mobility in two sulfide mine tailings from the Iberian Pyrite Belt (SW Spain). *Chem. Geol.*, 345, 119, 2013.
8. Lei, L. and Watkins, R., Acid drainage reassessment of mining tailings, Black Swan Nickel Mine, Kalgoorlie, Western Australia. *Appl. Geochem.*, 20, 661, 2005.
9. Álvarez-Valero, A.M., Sáez, R., Pérez-López, R., Delgado, J., Nieto, J.M., Evaluation of heavy metal bio-availability from the Almagrera pyrite-rich tailings dam (Iberian Pyrite Belt, SW Spain) based on a sequential extraction procedure. *J. Geochem. Explor.*, 102, 87, 2009.
10. Carbone, C., Di Benedetto, F., Marescotti, P., Martinelli, A., Sangregorio, C., Cipriani, C., Lucchetti, G., Romanelli, M., Genetic evolution of nanocrystalline Fe oxide and oxyhydroxide assemblages from the Libiola Mine (eastern Liguria, Italy): Structural and microstructural investigations. *Eur. J. Min.*, 17, 785, 2005.
11. Jambor, J.L., Mineralogy of sulfide-rich tailings and their oxidation products, in: *Environmental Geochemistry of Sulfide Mineral Wastes*, Short Course, vol. 22, J.L. Jambor and D.W. Blowes (Eds.), pp. 59–102, Mineralogical Association of Canada, 1994.
12. Carbone, C., Marescotti, P., Lucchetti, G., Martinelli, A., Basso, R., Cauzid, J., Migration of selected elements of environmental concern from unaltered pyrite-rich mineralizations to Fe-rich alteration crusts. *J. Geochem. Explor.*, 114, 109, 2012.
13. Bobos, I., Durães, N., Noronha, F., Mineralogy and geochemistry of mill tailings impoundments from Algares (Aljustrel), Portugal: implications for acid sulfate mine waters formation. *J. Geochem. Explor.*, 88, 1, 2006.
14. Gunsinger, M.R., Ptacek, C.J., Blowes, D.W., Jambor, J.L., Moncur, M.C., Mechanisms controlling acid neutralization and metal mobility within a Ni-rich tailings impoundment. *Appl. Geochem.*, 21, 1301, 2006.
15. Parandoush, K., Atapour, H., Riseh, M.A., Geochemical signatures of waste rocks around Sarcheshmeh porphyry copper mine dumps, southeastern

Iran: Implications for exploration, economic by-products and the environment. *J. Geochem. Explor.*, 199, 31, 2019.
16. Carrillo-Chavez, A., Salas-Megchun, E., Levresse, G., Munoz-Torres, C., Perez-Arvizu, O., Gerke, T., Geochemistry and mineralogy of mine-waste material from a "skarn-type" deposit in central Mexico: Modeling geochemical controls of metals in the surface environment. *J Geochem. Explor.*, 144, 28, 2014.
17. Andersson, M., Finne, T.E., Jensen, L.K., Eggen, O.A., Geochemistry of a copper mine tailings deposit in Repparfjorden, northern Norway. *Sci. Total Environ.*, 644, 1219, 2018.
18. Pasquini, A.I., Campodonico, V.A., Rouzaut, S., Giampaoli, V., Geochemistry of a soil catena developed from loess deposits in a semiarid environment, Sierra Chica de Córdoba, central Argentina. *Geoderma*, 295, 53, 2017.
19. Biswas, A., Hendry, M.J., Essilfie-Dughan, J., Geochemistry of arsenic in low sulfide-high carbonate coal waste rock, Elk Valley, British Columbia, Canada. *Sci. Total Environ.*, 579, 396, 2017.
20. Lieberman, R.N., Izquierdo, M., Córdoba, P., Palmerola, N.M., Querol, X., de la Campa, A.M.S., Font, O., Cohen, H., Knop, Y., Torres-Sánchez, R., Sánchez-Rodas, D., The geochemical evolution of brines from phosphogypsum deposits in Huelva (SW Spain) and its environmental implications. *Sci. Total Environ.*, 700, 134444, 2020.
21. Nehra, S., Raghav, S., Kumar, D., Biomaterial functionalized cerium nanocomposite for removal of fluoride using central composite design optimization study. *Environ. Pollut.*, 258, 113773, 2020.
22. Nehra, S., Manjula, N., Kumar, D., Hydrothermally shape-controlled synthesis of TiO_2/Graphene for fluoride adsorption studies. *J. Chem. Eng. Data*, 64, 5373, 2019.
23. Nehra, S., Raghav, S., Kumar, D., Rod–shaped Ca Zn@Chitin composite for fluoride removal studies by adsorption and statistical experiments. *Environ. Nanotechnol. Monit. Manag.*, 12, 100264, 2019.
24. Nehra, S., Raghav, S., Manjula, N., Kumar, D., Trimetallic oxide entrapped in alginate polymeric matrix employed for adsorption studies of fluoride. *Surf. Interfaces*, 13, 112–132, 2018.
25. Raghav, S., Nehra, S., Kumar, D., Cubical–shaped rods of pectin–hydroxyapatite composite for adsorption studies of fluoride by statistical method and adsorption experiments. *ACS Omega*, 3, 9675, 2018.
26. Raghav, S., Nehra, S., Kumar, D., Biopolymer scaffold of pectin and alginate for the application of health hazardous fluoride removal studies by equilibrium adsorption, kinetics and thermodynamics. *J. Mol. Liq.*, 284, 203, 2019.
27. Raghav, S., Nehra, S., Kumar, D., Adsorptive removal studies of fluoride in aqueous system by bimetallic oxide incorporated in cellulose. *Process Saf. Environ.*, 127, 211, 2019.
28. Benoit, B.B., Dann, E., Paullay, J.A., Fractal character of fracture surfaces of metals. *Nature*, 308, 721, 1984.

29. Paggi, M. and Barber, J.R., Contact conductance of rough surfaces composed of modified RMD patches. *Int. J. Heat Mass Transf.*, 54, 4664, 2011.
30. Goerke, D. and Willner, K., Normal contact of fractal surfaces—experimental and numerical investigations. *Wear*, 264, 589, 2008.
31. Carpinteri, A., Cornetti, P., Puzzi, S., Size effect upon grained materials tensile strength: The increase of the statistical dispersion at the smaller scales. *Theor. Appl. Fract. Mec.*, 44, 192, 2005.
32. Carpinteri, A., Cornetti, P., Puzzi, S., Size effect upon grained materials tensile strength: the increase of the statistical dispersion at the smaller scales. *Theor. Appl. Fract. Mec.*, 44, 192, 2005.
33. Sapna, K.D., Role of new generation technology to remediate environmental pollution, in: *Nanocomposites for Pollution Control*, A.K. Mishra (Ed.), pp. 1–40, Jenny Stanford Publishing, New York, 2018.
34. Sapna, K.D., Biodegradable polymer based nanoadsorbents for environment remediation, in: *New Polymer Nanocomposites for Environmental Remediation*, A.K. Mishra (Ed.), pp. 261–278, Elsevier, 2018.
35. Sapna, K.D., Carbon nanomaterials-based nanocomposite as emerging field for pollution control, in: *Nanotechnology in Environmental Science*, A.K. Mishra (Ed.), pp. 125–152, Wiley–VCH Publisher, USA, 2018.
36. Sapna, Sharma, R., Kumar, D., Nanomaterials as an emerging opportunity to procure safe drinking water using bio–polymeric materials, in: *Biopolymers: Structure, Performance and Applications*, A.K. Mishra (Ed.), pp. 67–94, NOVA Science Publishers, 2018.
37. Sapna, Sharma, R., Kumar, D., Chitosan based membranes for wastewater desalination and heavy metal detoxification, in: *Nanoscale Materials in Water Purification*, D.A. Gopakumar (Ed.), pp. 799–814, Elsevier, 2019.
38. Sharma, R., Sapna, Dhillon, A., Kumar, D., Biosorbents from agricultural by–products: updates after 2000s, in: *In Bio & Nanosorbent from Natural Resources*, A.K. Mishra (Ed.), pp. 1–20, Springer, Cham, 2018.
39. Sapna, Raghav, S., Kumar, D., The rising environmental and human health impacts of plastic pollution, in: *Review Environmental implications of incineration of municipal solid waste and ash disposal*, l, S.M. Zuber (Ed.), pp. 59–79, IGI global, 2020.
40. Khurana, P., Thatai, S., Sapna, Kumar, D., Destruction of recalcitrant nanomaterials contaminants in industrial wastewater, in: *Emerging and Nanomaterial Contaminants in Wastewater Advanced Treatment Technologies*, A.K. Mishra (Ed.), pp. 137–158, Elsevier, 2019.
41. Morales-Simfors, N., Bundschuh, J., Herath, I., Inguaggiato, C., Caselli, A.T., Tapia, J., Choquehuayta, F.E.A., Armienta, M.A., Ormachea, M., Joseph, E., López, D.L., Arsenic in Latin America: A critical overview on the geochemistry of arsenic originating from geothermal features and volcanic emissions for solving its environmental consequences. *Sci. Total Environ.*, 716, 135564, 2019.

42. Zouhri, L., El Amari, K., Marier, D., Benkaddour, A., Hibti, M., Bacteriological and geochemical features of the groundwater resources: Kettara abandoned mine (Morocco). *Environ. Pollut.*, 252, 1698, 2019.
43. Battioui, M., Bouzahzah, H., Benzaazoua, M., Hakkou, R., Sbaa, M., Column kinetic tests assessing geochemical behavior of mine wastes in the jerada coal district (morocco). *Mine Water Environ.*, 35, 97, 2016.
44. Wilson, D., Amos, R.T., Blowes, D.W., Langman, J.B., Ptacek, C.J., Smith, L., Sego, D.C., Diavik waste rock project: A conceptual model for temperature and sulfide-content dependent geochemical evolution of waste rock–Laboratory scale. *Appl. Geochem.*, 89, 160, 2018.
45. Stumbea, D., Chicoş, M.M., Nica, V., Effects of waste deposit geometry on the mineralogical and geochemical composition of mine tailings. *J. Hazard. Mater.*, 368, 496, 2019.
46. Shahhosseini, M., Ardejani, F.D., Amini, M., Ebrahimi, L., Poorkani, A.M., Environmental geochemistry of As and Pb in a copper low-grade dump, Miduk copper mine, Kerman province, SE Iran. *J. Geochem. Explor.*, 198, 54–70, 2019.
47. Zhang, S., Liu, G., Yuan, Z., Environmental geochemistry of heavy metals in the groundwater of coal mining areas: A case study in Dingji coal mine, Huainan Coalfield, China. *Environ. Forensics*, 20, 265–274, 2019.
48. Sharafi, A., Ardejani, F.D., Rezaei, B., Sargheini, J., Environmental geochemistry of near-neutral waters and mineralogy of zinc and lead at the Angouran non-sulphide zinc mine, NW Iran. *J. Geochem. Explor.*, 186, 7793, 2018.
49. Khorasanipour, M. and Jafari, Z., Environmental geochemistry of rare earth elements in Cu-porphyry mine tailings in the semiarid climate conditions of Sarcheshmeh mine in southeastern Iran. *Chem. Geol.*, 477, 58–72, 2018.
50. van der Ent, A. and Edraki, M., Environmental geochemistry of the abandoned Mamut copper mine (Sabah) Malaysia. *Environ. Geochem. Health*, 40, 189–207, 2018.
51. Tlil, H., Souissi, R., Souissi, F., Lattanzi, P., Podda, F., Concas, S., Ardau, C., Cidu, R., Environmental mineralogy and geochemistry of Pb–Zn mine wastes, Northern Tunisia. *Rend. Lincei*, 28, 133–141, 2017.
52. Essilfie-Dughan, J., Hendry, M.J., Dynes, J.J., Hu, Y., Biswas, A., Barbour, S.L., Day, S., Geochemical and mineralogical characterization of sulfur and iron in coal waste rock, Elk Valley, British Columbia, Canada. *Sci. Total Environ.*, 586, 753–769, 2017.
53. Li, L., Tu, H., Zhang, S., Wu, L., Wu, M., Tang, Y., Wu, P., Geochemical behaviors of antimony in mining-affected water environment (Southwest China). *Environ. Geochem. Health*, 41, 2397–2411, 2019.
54. Aravinthasamy, P., Karunanidhi, D., Subramani, T., Srinivasamoorthy, K., Anand, B., Geochemical evaluation of fluoride contamination in groundwater from Shanmuganadhi River basin, South India: Implication on human health. *Environ. Geochem. Health*, 42, 1937–1963, 2020.

8

Medical Geochemistry

Hosam M. Saleh* and Amal I. Hassan

*Radioisotope Department, Nuclear Research Center,
Atomic Energy Authority, Dokki, Giza, Egypt*

Abstract

Several diseases can generate many quandaries for medical societies. Accordingly, new categories of medical sciences may contribute to addressing a range of those health restrictions. Medical geology is a rapidly developing branch with the remarkable ability to serve medical communities to address a massive scale of environmental health issues. Medical geology can be considered an integral part of environmental health to deal with the effects of geological processes and materials on human and animal health. Medical geology aims to address health problems caused by geological materials and processes such as trace elements, rocks, minerals, dust, water, volcanic eruptions, and earthquakes. This branch has accurately defined the correlation between geological elements and human health for centuries. Among the environmental health problems that geologists work with the medical community are exposure to toxic levels of basic insignificant elements such as arsenic and mercury, exposure to dust and radioactivity, presence of organic compounds in drinking water, gases emitted from volcanoes, and so on. Geologists have developed a set of tools and databases that the environmental health community can use to treat pesticide-borne diseases. These scientists also helped establish mechanisms to clarify the spread of pollution in surface and ground water, as well as industrial pollution and occupational health problems. In this chapter, we will focus on the benefits of medical geochemistry as a very important branch of geochemistry, and we will also look at the future of this science and what it can do to humanity to protect the environment and human health from the dangers of pollution.

Keywords: Geochemistry, isotopes, human health, pollution, medical geology

**Corresponding author*: hosam.saleh@eaea.org.eg; hosamsaleh70@yahoo.com

Inamuddin, Mohd Imran Ahamed, Rajender Boddula and Tariq Altalhi (eds.) Geochemistry: Concepts and Applications, (127–148) © 2021 Scrivener Publishing LLC

8.1 Introduction

Geochemistry is concerned with studying the composition of the Earth and other planets, the chemical processes and reactions that control the composition of rocks, water, and soil, as well as the cycle of matter and energy that transports the chemical components of the planet in time and space and considers the distribution and movement of elements in different areas of the Earth (crust, scarves, and water bodies) and minerals in order to determine the underlying system of distribution and movement [1]. In 1924, Clarke defined the concept of geochemistry and all geological studies as those related to chemical change [1]. One of the primary goals in geochemistry is to estimate the abundance of elements in the Earth's crust and the main layers of rocks and minerals. Therefore, the application of geochemistry techniques in developing critical solutions to geological and environmental problems is a very effective way to interpret analytical results [2].

Geologists have confirmed that the Earth is the only planet in our solar system known for its continental crust. This crust remained a source of controversy for a long time, due to its structural heterogeneity, which impedes the direct testing of morphology models [3]. Keller and his colleagues have confirmed that the majority of igneous rocks in the continental crust had formed by the hardening of lava or volcanic magma that are formed gradually by the crystallization of the molten material [4]. Daley's theory assumed that the relationship between the concentration of oxides, such as magnesium oxide and silica, should be linear. The long-standing issue in Bowen's hypothesis is that the remnants of ferromagnetic crystals must have accumulated in the crust over time, forming cumulative rocks [5]. Numerous evidences, including data from geological samples and numerical simulations, support a model in which the residues of iron ore and magnesium crystals possess a greater density than the upper mantle layer, separating them from the base of the continental crust [6]. Based on the researchers' extensive calculations, these differences are consistent with the expected cumulative structures and prove that the cumulative iron ore and magnesium crystals are present in underground rocks, but not in abundance. The researchers observed that most of the continental underground rock formations have a higher content of silica than expected, compared with the accumulation of iron ore and magnesium. Keller and colleagues also concluded that underground rocks formed in fractured continental environments have characteristics consistent with the idea that water exists during their formation, where it was previously believed that these environments were drying out [7]. The presence of water can explain the discontinuation and crystallization of magma in the continental crust, rather than its emergence to the surface [7]. Water-containing magma can reach the temperature at which the crystallization process begins to coincide with its rise, while dry magma does not. Ultimately, the floating continental crust possesses geochemical patterns that are consistent with the idea of formation with small amounts of water [8]. The analytical evidence for the current data set suggests that the composition of the rocks' group further than one million years is not fundamentally different from the younger group. Many old rocks were lost during recycling such as erosion, and those that survived recycling may not represent the diversity that existed previously [9].

8.2 The Evolution of Geochemistry

In the early 20th century, the two hypotheses that emerged in leading experimental geochemistry laboratories competed and influenced almost all models of continental crust formation. In 1915, Bowen noted that the continental shell was developed through a gradual crystallization, and magma distilled from the ground mantle layer. This theory assumes that minerals that crystallize first are those with a higher proportion of iron and magnesium. Geochemistry lacks a central concept and is, unlike ecology, which has a beneficial concept of an ecosystem as a focus and which may be linked to local, regional, or global environmental studies. Possibly, the closest way to a central concept is the geochemical cycle [10, 11].

The field of geochemistry has evolved since the early days of Bowen and Dali to include observations ranging from the microscopic scale to the universe. Combining geochemical investigations with other sciences helps to ensure that the continental crust has played a significant role in the rise of oxygen in the Earth's atmosphere [12].

8.3 This Science has Expanded Considerably to Become Distinct Branches

General geochemistry: It deals with the study of laws that control the movement of chemical elements in geological processes such as weathering, transport, sedimentation, and transformation and form various useful ores [13]. This branch of geochemistry has solved many of the theoretical questions of practical application, such as the mechanism of transmission of the elements that make up non-soluble compounds in water, the mechanism of the arrangement processes and its causes, and many laws governing the distribution of useful ore deposits [13].

Geochemistry of the elements: It is concerned with the study of the distribution of chemical elements, their forms of presence, and the rates of their concentrations in metals, rocks, and the covers of the earth [14]. This section has revealed many facts about the distribution of chemical elements. The most prevalent elements in the Earth's crust are light elements with an atomic weight of less than 29, and the elements of oxygen, hydrogen, silicon, and aluminum are the most prevalent and important in the formation of the Earth's crust and the covers of the earth [15]. The first is about 50% in the composition of the Earth's crust, and its importance for life is known, and the first and second make up the water, and of these four elements combined consists of rocks that make up more than half the size of sedimentary rocks, and that elements with numbers and weights of atomic are more widespread than others [1, 16]. However, there are many issues that require a solution, such as the high variability in the concentrations of chemical elements, and the presence of scarcity elements encountered in the case of scattering among other metals such as rubidium and others [17].

Biogeochemistry: It properly examines the dispensation of chemical elements and their concentrations in organic matter and the performance of organisms in the movement of ingredients and their distribution in Earth's various layers. Although the weight of living organisms is estimated at 0.01% of the weight of the Earth's crust, they have a critical role in geological processes, called biochemical processes [18]. For example, iodine is obtained mainly from some marine plants that extract it from seawater and concentrate it in their tissues in proportions hundreds of times more than in water. Photosynthesis and the release of oxygen by the influence of solar energy also store this energy in the form of fossil fuels (coal, oil, and gas) [19].

An important topic in this science is the study of biogeochemical districts, areas where the concentration of chemical elements changes from their overall rates of increase or decrease. This is reflected in organisms by exposure to disease or the emergence of some changes [20]. At least 30 chemical elements were found to be involved in the formation of these provinces. Among them are the iodine-poor districts, which are found in highland areas, the interior of continents, etc., the fluoride-poor districts that cause tooth decay, and the fluorine-rich districts that are spread in volcanic areas and in areas where apatite rocks are spread and usually cause osteoporosis [21]. This type of studies is used in the treatment by adding the necessary elements to water, fertilizers, and animal feed.

Natural water geochemistry: It studies the role of water in the migration of chemical elements and their redistribution in the Earth's crust and land cover. The most important topics of this section are the study of the origin and chemical composition of water and the forms of the presence of chemical elements in it or the processes that take place between water and benthic sediments, and between them and the gas atmosphere [22]. As well as the study of sedimentation processes, the formation of salts, and prospecting for useful ores (hydrogeochemical methods). This science is divided into two main branches: seas and oceans geochemistry and continental waters geochemistry, also called hydrogeochemistry [23].

Isotope geochemistry: It studies the distribution of isotopes in the chemical constituents of natural bodies (rocks, metals, water, and organic matter). Studies have shown that the proportion of isotopes of the chemical element changes by the impact of geological processes and by studying these changes can solve many issues such as the origin of rocks, metals, and useful ores, and the temperature of their formation, and the source of chemical elements [24].

8.3.1 Cosmochemistry

It is interested in studying the distribution of chemical elements and their concentrations in celestial bodies, especially meteorites [25]. It also studies the effect of cosmic rays in the components of the Earth's envelopes, especially the gaseous atmosphere, which causes the formation of some elements such as carbon C_{14}. Cosmochemical studies are useful in scrutinizing information about the origins and evolution of the Earth, its shell formation, and the evolution of geological processes during Earth's long history. This, in turn, is useful in guiding the exploration of useful materials [25].

8.3.2 The Economic Importance of Geochemistry

The primary objective of geochemical studies is to detect useful ore deposits. They are particularly important because they detect deep reservoirs that are difficult or impossible to detect by other geological methods, especially after most of the shallow and shallow reservoirs have run out. In addition, geochemistry helps to solve many geological issues such as determining the absolute age, origin of rocks, metals, and ores, the conditions of their formation, and other important things that help guide the exploration of useful ores [26].

8.3.3 Analytical Geochemistry

In the applications of theoretical calculations and forecasts within geochemistry [27], most geochemical studies fall within the scope of analytical geochemistry and are concerned in one way or another with sample measurement. Geochemical analysis is a big and important topic in this field, which is measuring the quantities of elements, isotopes, minerals, and compounds present in rock, natural liquid, or natural gas [27]. Fields such as isotope geochemistry are dominated by techniques such as mass spectroscopy, despite their great diversity in the type of sample being analyzed and the preparation of machines in particular [28]. In fields such as sedimentary, fiery and metamorphic geochemistry, or other branches of geochemistry that are closely related to petroleum science, a variety of analytical techniques are often used whether the composition of rocks, glass chemistry, or minerals is important. Mineral analysis includes determining the chemical relationships between and inside mineral grains. Precise analytical techniques are necessary, and methods such as X-ray spectroscopy, ionic radiation techniques, and laser ablation procedures are usually used to analyze major and trace elements [29]. In addition to the minerals themselves, the analysis of fluid impurities (small drops of fluid that were trapped inside the crystals either during the initial growth of the solution or at a later stage) made a significant contribution to the development of modern theories of ore origin, reproduction, formation, oil migration, and accumulation, and our understanding of the importance of liquid stage in a wide range of geological processes [30]. Other specialized techniques for extracting information from minerals include fission path analysis, in which the tracks of physical damage arising from spontaneous fission of the original nucleus to the crystal network are measured. For other geological techniques, when the rate of occurrence of spontaneous fission is known, the accumulation of these fission paths can be used as a dating tool [31].

8.3.4 Geochemistry of Radioisotopes

Isotopic geochemistry is the comprehensive examination of the relative and absolute concentrations of elements and their isotopes in samples from the Earth and the solar system [31]. Such measurements provide a powerful tool for questioning a range of scientific problems ranging from the origins of terrestrial planets, past climate change, obsolete processes, and the source of elements in a variety of geological repositories [32]. It can be divided into the study of radioisotopes and stable isotopes. In addition to radiation degradation, a variety of processes (physical, chemical, kinetic, and equilibrium alike) have the ability to segment the isotopes of the element, which means that the elements develop with different isotopic signatures in already different geological and cosmic contexts [33]. Hence, the general instrument of isotope geochemistry is useful in several ways. For example, the use of known degradation rates of radioisotopes in geological techniques is known in time and widely in Earth and environmental sciences to determine the age and/or stratum location of samples or the rate of operations [34]. Differences in the ratios of stable isotopes within stratified columns, usually associated with magnetic information and biographical graphics, can also be used to connect the horizons of different stratum layers [34]. Stable isotopes as traces of global cycles have also made an important contribution to understanding processes such as rock and water cycles and changes in the Earth system over the geological time with implications for future anthropogenic climatic and environmental impacts [35].

Geochemistry utilizes theoretical fractions resulting from the influence of the neutron mass on the following; a) chemical bonding, b) diffusion, c) reaction rates, and d) the ability of molecules to transfer between energy states through stable isotopes [36]. These ultimate outcomes are remarkably small and typically appear in a considerable percentage or less to isotope divergence. Hydrogen is an exception with differences in the ratio of 2H/1H often 10%. Greatest of the locus in this field is on a few optical elements such as, H, C, N, O, and S. temperature operations at or near the Earth's surface [37]. Since a result of this temperature dependency, fragmentation of the isotopes of oxygen between the layer of water and the calcium carbonate of marine organisms provides useful Balinese thermometers [38] to, among other things, the detection of ice-age fluctuations [39, 40] among the largest Geochemical contributions to stable isotopes. In current years, this has been reinforced by considerations of hydrogen isotopes in the ice layers in Antarctica and Greenland [40]. Geothermal engineering of sulfur isotopes has been applied to the origins of sulfur ore deposits. The ultimate end of the fractionation independent of the isotopes of sulfur 3.3 billion years ago authoritative records the initial oxidation of the atmosphere [41]. Photosynthesis is the main cause of differences in the ratios of carbon isotopes, and all biocomposites are debilitated at 13°C, usually in the range of 15–25 or more per ml. That essential oil holds this isotope indication removes any doubt about its ultimate biological origin.

This sign itself is found in a subset of diamonds bearing a metal impurity deep cloak, which leads to the most surprising conclusion that the carbon in it is derived from organic matter carried in the inward mantle [42].

The compounds produced in 13°C are exhausted, and the remainder of the inorganic carbon in the system, such as the ocean. This process has been used to track both oceanic circulations [24, 43], and biological productivity and diversity over time, indicating the association of these to biochemical and tectonic bicycles [43]. The chemistry of stable isotopic geology has proven to be beneficial in archeology and paleontology. Carbon and nitrogen isotopes in bone collagen and apatite are widely used to identify food sources for peoples and animals. It has been distinguishing between marine diets and terrestrial species and the extent of dependence on C4 plants such as corn, sugar cane, and C3 plants, such as wheat, barley, etc. [44, 45]. Several exciting new developments in stable isotope chemistry have occurred over the past few decades [46]. One of these factors is the development of new analytical techniques that allow more accurate and on-site measurements on very small samples. Another development is the special complex theoretical analysis as the one mentioned in the previous paragraph. Another development is the extension of the geochemistry of stable isotopes from a few light elements to almost every multi-isotope component in the periodic table, including most transport minerals as well as very heavy metals such as U and Tl [47]. A further approach is the evolution of mass isotopes: the isotope analysis of pairs of elements, such as C and O, in individual theoretical types or isotope isotopes. The noble gas isotopes (NGIs) are of particular importance because many of them are generated either through radioactive decay (RD) or secondary nuclear reactions resulting from RD or radiations of cosmic or that have witnessed mass-dependent fragmentation [36, 48]. The geochemistry of NGIs has been applied to various studies such as crust fluids, ocean circulation, and corrosion rates. Xenon isotopes resulting from the extinction of radionuclide decay ^{129}I and Plutonium 244 provide the early history of Earth and its atmosphere [49].

8.3.5 Medical Geochemistry and Human Health

Medicinal chemistry is the dialectical link between chemical elements, human health, and the ecosystem (geochemistry). The trace elements are found in very minimal amounts in the vital tissues and body fluids [50]. Despite their rarity, they play an important role in the body's vital functions especially as it relates to biomolecules such as enzymes. Most of the trace elements are transitional elements, and most of them in the fourth cycle of the periodic table, except for boron, selenium, molybdenum, iodine, and tungsten, they are distributed elsewhere in the periodic table [51]. All of these elements have important vital functions. Each of the scarce chemical elements is present in the human body with a certain range of concentration, and it is not required to be of homogeneous distribution, but most of these elements are concentrated in a specific tissue which is the accumulation of existence [51]. Any increase or decrease in the normal range of any chemical element in the human body results in symptoms of medical conditions.

The source of all these elements is the earth with its minerals, rocks, and water. The cycle of elements in nature plays a major role in the redistribution of these elements in different geological circles. The biological envelope of the Earth, meaning plants and animals, is a storehouse of organic materials interacting with the environment and the surrounding environment [52]. The science that clarifies the concentrations of elements in the Earth, both quantitatively and qualitatively, is known as chemistry. The human body will obtain these elements in different ways, but its primary source is the Earth. Since human health is linked to specific concentrations of each component in the human body, then from here it highlights the role of medical geochemistry and its role in physiotherapy and maintaining general health.

For example, the element of vanadium in geological terms, a rare element in the Earth's crust, as few quantities of it were found in meteorites that come from the planets of the solar system except Earth [53].

Vanadium is not found in a free form. Rather, it is found co-exist with lead in vanadium, and it is associated with Uranium in the radioactive carnotite metal $K_2(UO_2)_2(VO_4)_2 \cdot 3H_2O$ and with Sulfur in the form of Patronite (VS_4, $V(4^+)(S_2(2^-))_2$). It is found with iron and titanium ores within titanium magnetite, it is also found in fire rocks and accompanies hydrocarbons. Vanadium is found in seawater in the ratio of 2 ppm, and in fresh water, it is much less [54]. The processes of metabolism that occur in the aqueous media in a human body rich in vanadium. Vanadium ions, vanadinite (VO_3^-), and vanadyl (VO^{+2}) can bind to molecules of physiological importance in the body [54]. As there are positive and negative oxides, which are in both cases oxidizing agents that have the ability to interact through a two-stage electron transmission, where the vanadium interacts here as a carrier of the oxygen atom (electron donor). Vanadium joins with sulfur to form negative roots like (VS_4^{-3}), and positive roots like VS^{+2} and $VSSH^+$. The vanadium is concentrated in the tissues and fluids of the human body in small quantities, since the adult human body contains about 100 ppm of vanadium, which is concentrated in the tissues of muscles, fluids, and bones, and it is a cumulative component, which accumulates in the heart muscle, spleen, thyroid gland, lung, and kidneys [55]. The intestine absorbs very scarce concentrations of vanadium quantities entering the body [56]. Therefore, a liter of water does not meet the daily needs of vanadium. Vanadium plays a role in the process of digesting and breaking down carbohydrates, and it has a role in the functioning of the circulatory system. It regulates the work of potassium, sodium triphosphate, and other enzymes [57]. Additionally, it activates the hepatocytes by obtaining the oxygen. It has other functions related to eyesight, liver, kidneys, heart, nervous system, and insulin regulation in the blood. Therefore, vanadium is considered to be like insulin in its effectiveness to metabolize blood sugar. Accordingly, the low concentration of this element in the body causes arteriosclerosis, diabetes, sagging muscles, incomplete growth, and many other pathological symptoms [58]. Also, the increased concentration of vanadium in the body causes the appearance of poisoning in the case of high concentrations, depression, inflammation of the mucous membranes of the eye and the digestive system, asphyxia, and deficiency in immunity through the breakdown of white blood cells and anemia. Hence, the urgent need is to know its natural resources and how to maintain its normal concentration within the human body. The most important nutritional sources of vanadium are black pepper, dill seeds, and mushrooms. Cobalt is a transitional chemical element and possesses magnetic properties. Its normal distribution ranges from 20 to 30 ppm [59]. The main source is a variety of geological sources, and the most important sources of this trace element are the deposits of manganese nodes on the bottom of the ocean, that emerge from the volcanic activity in the seabed and ocean's bottom. Cobalt accompanies sulfide and arsenide ores and is also a separate mineral called cobaltite. All of these are natural geological sources. The elements are distributed over the Earth's crust by various geological processes such as dissolution, transport, and then re-distribution and sedimentation [60].

Cobalt is essential to the health and safety of the functional functioning of the human body in several ways. As it is closely related to proteins in the body, it is found in vitamin B12 and is considered a catalyst in this vitamin. This vitamin includes many reactions, including reducing ribose to deoxyribose and relocating methyl groups. Cobalt contributes to the metabolism of carbohydrates and proteins [60]. It helps in the production of amino acids, the creation of DNA and supports the immune system, the nervous and the digestive system in performing the vital functions, and it is responsible for monitoring the work of cells, growth, and the work of red blood cells and bone marrow, and activates some enzymes that need adults of 1.5 μg. Vitamin B12 ensures the necessary cobalt, which is mostly concentrated in bones and muscles. The heart contains the largest amount of this element, as 3 mg is entered into the body daily, and 85% of it is excreted with urine [61]. The body gets cobalt from vegetable food, which in turn absorbs from the soil (geological medium). Generally, cobalt is found in vegetables such as broccoli, spinach, fish, and grains, and white and red meat, legumes, seafood, and shellfish. Each 100 g of leafy vegetables contains 60–20 mg of cobalt, and every 100 g of meat contains 25–15 mg of cobalt [61].

High concentrations of cobalt in the body adversely affect the safety and functioning of the heart and may affect fertility. So, it may have had a prominent role in the processes of mass extinction of some species and animal species throughout the geological ages. As studies and experiments have shown, the deficiency of cobalt in the bodies of biology is a good example that links the health of an organism with the chemistry of the environment [62]. Hence, it turns out that the chemical balance of the health of living organisms is very delicate, as changes in very fine amounts of trace elements can have major impacts on public health and safety [63].

8.3.6 Environmental Health and Safety

There are over two million chemicals known to date, and every year, over a thousand chemicals are discovered by the chemical factories. Unfortunately, hundreds of these chemicals are used commercially and do not know enough information about the effect of most of these chemicals on health [64]. Also, environmental pollution occurs as a result of airborne particles that come from diverse sources and processes. Thus, omnipresent in the lower troposphere, there is a mixture of solid particles of different origins. It is worth noting that between 3.4 and 12 billion tons of particles are released globally every year. These releasing particles are manufactured of inorganic and organic composites from anthropogenic and natural sources [65]. Airborne molecules of natural origin far outperform anthropogenic, with airborne geological and biogeological molecules representing for more than 90% of the total mass of molecules [66].

The hazard of health difficulties associated with susceptibility to metals and metalloids in airborne particles was the goal of a large number of toxicological and epidemiological studies [65]. There is a list of diseases believed to be due to the presence of chemicals in the environment. These diseases include lung damage and emphysema related to lead air pollution in paints or added to gasoline, heart disease and carbon monoxide, permanent nerve damage and mercury, excess in copper that causes liver cell poisoning, aluminum build-up in the brain, and Alzheimer's disease in addition to lots of Chemicals potentially linked to cancer [67]. It is clear from the above that the industrialized world has introduced many chemicals and found their way into the environment to cause pollution that harms human (Figure 8.1). Metals can raise the acidity of the blood, which in shift affects the balance of calcium and thus stimulates the release of calcium from the bone. Moreover, metals and metalloids can arouse or aggravate allergic reactions and compete with other essential trace minerals for biochemical bond sites. Toxic metals and metalloids tend to accumulate *in vivo* if there are no mechanisms to remove them or get rid of those toxins. The main ways of excretion are the kidneys and liver. While the biological fluids such as saliva, tears, and urines, as well as tissues such as nails and hair are less important in terms of reducing overall body burden [68].

Although carbon dioxide (CO_2) has a prominent role in photosynthesis, it is sometimes related to pollution. The high levels of CO_2 in the atmosphere lead to an increase in the acidity in ocean water and have harmful effects on marine ecosystems. Many chemicals pollute the soil and water, and some of these compounds settle in the place that has been contaminated for a long time without any chemical changes [69]. There are other compounds with short codification, as they are chemically altered by heat, humidity, light reactions, microbes, and other factors. Contaminated soil and water chemicals include pesticides that are widely used in agricultural purposes for pest resistance, but at the same time, they harm the crops. These pesticides are usually used through spraying, where they mix with air and then fall on the soil and water, including those that settle in the

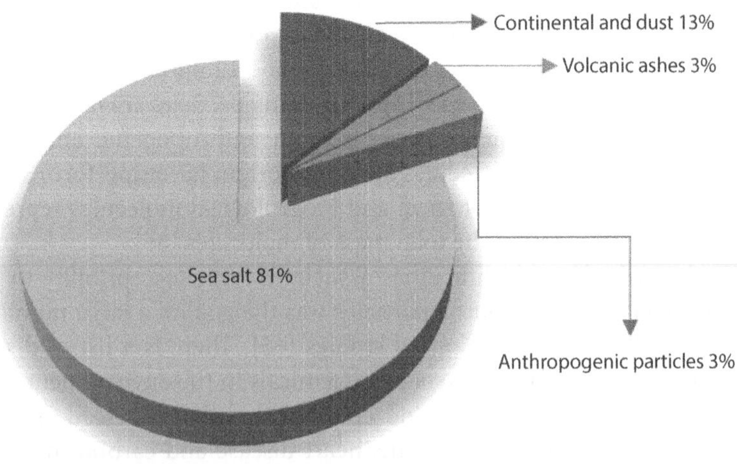

Figure 8.1 The percentages of particles loaded in the atmosphere from natural and human sources. Modified from [66].

contaminated place for a long time, such as chlorine hydrocarbons (dichlorodiphenyltrichloroethane (DDT), Aldrin, heptachlor, Linden, Lindane, and toxaphene) [70]. These pesticides are characterized by slow chemically degradation in soil and water by microbes to a greater extent by less chemical and optical reactions. Because these compounds settle in soil and water for a long time, they are considered one of the most dangerous pesticides for plants, birds, animals, and aquatic organisms. Some compounds settle for a medium period and include weed killers, such as Triazine compounds and phenyl urea compounds, and these compounds are chemically degraded in soil and water in a period less than the previous group, due to the influence of chemical and light reactions, so these compounds are considered less dangerous than the previous group on animals, birds, aquatic organisms, and plants [71]. Some compounds settle before being decomposed chemically for short periods ranging from several hours to several weeks or months. These compounds include herbicides from the phenyl carbamate group and fungicides from derivatives of dithiocarbamate in addition to organic phosphorus compounds and carbamate compounds that are used as insecticides [71]. These compounds, even if they are chemically degraded in a short period, but some of them may represent a danger to humans and animals, as some of them, such as dithiocarbamate compounds, may be transformed in the soil into carcinogenic substances [72].

The photosynthesis is considered one of the most important processes in green plants, and any chemical intervention that works to inhibit or stop any enzymatic reaction within the complex reactions of photosynthesis means that the plant stops obtaining the necessary food [73]. Chemicals that can interfere with these processes are not necessarily one of the most efficient weed killers, because most of the enzymes responsible for forming sugar in chloroplasts are also present in different cell cytoplasm, where they can be catalysts for adverse reactions, so the plant can get energy from stored materials, during breathing operations, until it regains its ability to perform the process of photosynthesis again, which requires that the mate rial be inhibited enough to discourage the formation of sugar and to discourage the burning process in breathing operations simultaneously [73]. It has already been found that many active compounds such as herbicides are those that interfere with the process of photosynthesis by affecting the sensitivity process of capturing light energy by chlorophyll and transferring that energy to the receiving molecules in them in the form of chemical energy [74]. The chlorophyll molecules appear to be tightly stacked, allowing chemical energy to pass through many of them. There is information available that the triazine compounds and the substituted urea compounds impede the process of photosynthesis by locking the circuit and the absence of the continued transfer of light energy to the receiving molecules. It was found that one molecule of herbicide against 600 chlorophyll could stop the photosynthesis process [75].

There are many chemical compounds other than pesticides that may pollute the soil and water, including industrial waste and sewage. Also, the purification of drinking water using chlorine leads to the formation of chlorine hydrocarbon compounds, which are considered one of the most important water pollutants [76]. There are also hydrogen carbon compounds that form in water mainly using chlorine in water purification chloroform and bromoform compounds. The danger of these compounds is that they may cause colon, rectal, and bladder cancer [76], as well as aromatic hydrocarbon compounds and examples of these compounds are polychlorinated biphenyls, which are compounds used in some industries such as papermaking [77]. These compounds are considered one of the most dangerous soil and water pollutants, as they represent harm to humans, animals, and aquatic organisms. There are also chlorophenol compounds that are used to preserve wood, and some of them are used in the manufacture of soap and deodorants. These compounds are considered one of the most dangerous water pollutants, as they cause health damages to humans and animals [78].

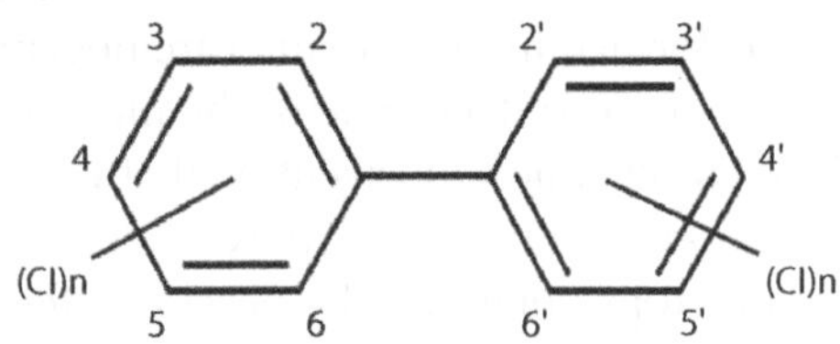

Polychlorinated biphenyls (PCBs)

PCBs with a dioxin-like activity has been known to cause a variety of teratogenic effects in animals. Exposure to PCBs causes hearing loss and symptoms similar to hypothyroidism in mice and rats. Besides, some human studies have reported a link between environmental levels of PCBs and the immune response, malignant melanoma, and rare liver cancers [79].

Heavy metals, such as mercury, lead, arsenic, cadmium, and selenium, are among the most dangerous materials that contaminate soil and water, and the most important sources of this pollution are planting waste, metal smelting, coal combustion, car exhaust, and pesticides that contain arsenic [80].

Inorganic compounds such as nitrates, phosphates, and fluoride are among the most important materials that pollute soil and water [81]. Nitrate and nitrite compounds are used as fertilizer, as a result of the mixing of animal waste, lead to soil and water pollution. The contaminated water or foods with nitrates induces high levels of oxidized hemoglobin in the blood, which leads to the inability of hemoglobin to deliver oxygen to the tissues of the body [82]. One of the most dangerous effects of nitrite compounds is that they interact with the amino substances in the food to turn into a toxic substance called nitrosamine. This substance causes injuries to the liver, lung, and nervous system [83]. It is also considered one of the substances that cause cancer and fetal abnormalities [83].

Soil and water are contaminated with phosphate compounds following the use of industrial fertilizers that contain these compounds. Sources of this pollution also include detergents that contain phosphate compounds and mix with soil and water by treating sewage. The increase in the percentage of phosphates in lakes and ponds leads to the growth of algae on the surface of the water, which affects the purity of the water [84].

Underground water collects under the outer crust of the earth, and this water is considered one of the most important water sources that countries give the highest attention to preserving them from environmental pollution. Environmental pollution and the indiscriminate use of groundwater threaten the world's groundwater resources. The pollution of groundwater evoked from the waste of factories, oil pipelines, industrial fertilizers, pesticides, and radioactive materials [85]. The problem of ocean and seawater pollution is one of the most serious environmental problems globally. The reasons for this pollution are due to the dumping of ship waste from petroleum and other chemicals. In addition to contamination of water by the waste of factories that contain organic compounds and toxic heavy metals, these pollutants represent the greatest danger to aquatic organisms, as they lead to a deterioration in their growth and reproduction [86].

One of the reasons for the increase in air pollution is the accumulation of toxic wastes that contributed to the emission of toxic gases and radioactive pollution into the atmosphere [87]. Some of these sources are considered natural that are produced from volcanoes and what is penetrated the atmosphere as a result of the ozone hole, which causes harm to the biosystem [87].

Air pollution varies from place to place according to wind speed and weather conditions. For example, nitrogen oxides interact with hydrocarbons in the presence of sunlight under special weather conditions, often in the summer to produce toxic chemicals such as pyroxene tetra acetyl and ozone gas [88]. Animal studies have found that exposure to nitrogen dioxide concentrations less 1.880 $\mu g/m^3$ than causes adverse and irreversible lung effects and biochemical changes ranging from several weeks to months [89].

Climate and daily fluctuations in the atmosphere are important determinants of conditions in a region, such as vegetation growth and energy use. With industrial progress, greenhouse gases accumulate in the atmosphere, causing global warming. Greenhouse gases include carbon dioxide (55%), nitrogen oxides (6%), chlorofluorocarbons (24%), and methane (15%) [90].

8.4 Conclusion

Medical geology is a dynamic new system that combines Earth Sciences, Biomedicine, and Community General Health to solve and address many environmental health problems. Medical geology is one of the latest complexes and complex science that requires the contribution and participation of many other scientific disciplines. Most of the chemical elements present in nature in the form of metallic and non-metallic minerals that enter our life system directly or indirectly and increase their concentration in the human body lead to a harmful effect on human health. Also, increasing the concentration of these elements in the soil leads to damage to plants and consequently affects animals in one way or another.

Through preliminary studies and surveys, it was found that some of these elements mainly affect our lives, while others, its presence or lack thereof, do not affect our health. The presence of toxic terrestrial elements or radioactive elements in nature was the result of natural factors, whether erosion or weathering of rocks or as a result of various activities that are human-induced that lead to the concentration of these elements in the human body either through food, drinking water, or breathing polluted air.

Pollution does not result from a specific sector or from one of the human activities. Rather, it includes all areas where people seek food, housing, clothing, and entertainment services. New challenges have been presented to the medical profession to focus seriously on an emerging field of medical geochemistry in cooperation with geologists and epidemiologists. Geological materials such as dust, soil, water, and toxic substances released from them can chemically interact with body fluids, thus these chemical reactions can play critical roles in toxicity. Besides the harmful influences of some geological substances, some clays have shown remarkably antimicrobial properties, particularly bacterial infection. Medical geochemistry can undoubtedly play a significant role in helping to protect and safe drinking water by correctly identifying sources, concentrations, and shapes of potential elements such as arsenic, mercury, and fluorine in natural waters.

References

1. Manahan, S., *Environmental chemistry*, CRC press, Taylor and Francis Group, 2017.
2. Tedesco, S.A., *Surface geochemistry in petroleum exploration*, Springer Science & Business Media, Berlin/Heidelberg, Germany, 2012.
3. Rauchfuss, H., *Chemical evolution and the origin of life*, Springer Science & Business Media, Berlin/Heidelberg, Germany, 2008.
4. De Boer, J.Z. and Sanders, D.T., *Volcanoes in human history: the far-reaching effects of major eruptions*, Princeton University Press, Princeton, New Jersey, 2012.
5. Albarède, F., Goldstein, S.L., Dautel, D., The neodymium isotopic composition of manganese nodules from the Southern and Indian oceans, the global oceanic neodymium budget, and their bearing on deep ocean circulation. *Geochim. Cosmochim. Acta*, 61, 6, 1277–1291, 1997.
6. Davis, G.H., Reynolds, S.J., Kluth, C.F., *Structural geology of rocks and regions*, John Wiley & Sons, United States, 2011.
7. Dohm, J.M. *et al.*, Recent geological and hydrological activity on Mars: the Tharsis/Elysium corridor. *Planet. Space Sci.*, 56, 7, 985–1013, 2008.
8. Tiab, D. and Donaldson, E.C., *Petrophysics: theory and practice of measuring reservoir rock and fluid transport properties*, Gulf professional publishing, Houston, Texas, United States, 2015.
9. Berner, R.A., GEOCARBSULF: a combined model for Phanerozoic atmospheric O2 and CO2. *Geochim. Cosmochim. Acta*, 70, 23, 5653–5664, 2006.
10. Fortescue, J.A.C., *Environmental geochemistry: A holistic approach*, vol. 35, Springer Science & Business Media, Berlin/Heidelberg, Germany, 2012.
11. Newman, P. and Jennings, I., *Cities as sustainable ecosystems: principles and practices*, Island Press, Washington, D.C., United States, 2012.
12. Helffrich, G.R. and Wood, B.J., The Earth's mantle. *Nature*, 412, 6846, 501, 2001.
13. Randive, K.R., *Elements of Geochemistry, Geochemical Exploration and Medical Geology*, Research Publishing Service, Singapore, 2013.
14. Kabata-Pendias, A. and Mukherjee, A.B., *Trace elements from soil to human*, Springer Science & Business Media, Berlin/Heidelberg, Germany, 2007.
15. Fyfe, W.S., *Fluids In The Earth's Crust: Their Significance In Metamorphic, Tectonic And Chemical Transport Process*, vol. 1, Elsevier, Amsterdam, Netherlands, 2012.
16. Press, F., Siever, R., Grotzinger, J., Jordan, T.H., *Understanding earth*, Macmillan, New York City, United States, 2004.
17. Brantley, S.L., White, T.S., White, A.F., Frontiers in exploration of the critical zone, National Science Foundation, Newark; Workshop on Frontiers in Exploration of the Critical Zone eng 24-26 Oct 2005 Newark, 2005.
18. Schlesinger, W.H. and Bernhardt, E.S., *Biogeochemistry: an analysis of global change*, Academic Press, 2013.

19. El Bassam, N., *Energy plant species: their use and impact on environment and development*, Routledge, 2013.
20. Funari, E., Manganelli, M., Sinisi, L., Impact of climate change on waterborne diseases. *Ann. Ist. Super. Sanita*, 48, 473–487, 2012.
21. Burkhardt, C., Borg, L.E., Brennecka, G.A., Shollenberger, Q.R., Dauphas, N., Kleine, T., A nucleosynthetic origin for the Earth's anomalous 142 Nd composition. *Nature*, 537, 7620, 394, 2016.
22. Bartram, J. and Ballance, R., *Water quality monitoring: a practical guide to the design and implementation of freshwater quality studies and monitoring programmes*, CRC Press, Boca Raton, Florida, U.S.A., 1996.
23. Guilbert, J.M. and Park Jr., C.F., *The geology of ore deposits*, Waveland Press, Long Grove, Illinois, 2007.
24. Sharp, Z., *Principles of stable isotope geochemistry*, Chapter 8, pp. 8–3, University of New Mexico, U.S.A., 2017.
25. McSween Jr., H.Y. and Huss, G.R., *Cosmochemistry*, Cambridge University Press, Cambridge, England, 2010.
26. Edwards, R., *Ore deposit geology and its influence on mineral exploration*, Springer Science & Business Media, Berlin/Heidelberg, Germany, 2012.
27. Labanowski, J.K. and Andzelm, J.W., *Density functional methods in chemistry*, Springer Science & Business Media, 2012.
28. Fry, B., *Stable isotope ecology*, vol. 521, Springer, Berlin/Heidelberg, Germany, 2006.
29. D'amore, J.J., Al-Abed, S.R., Scheckel, K.G., Ryan, J.A., Methods for speciation of metals in soils. *J. Environ. Qual.*, 34, 5, 1707–1745, 2005.
30. Godfrey, L.V. *et al.*, Hafnium and neodymium isotope variations in NE Atlantic seawater. *Geochem. Geophy. Geosy.*, 10, 8, 1–13, 2009.
31. Chakrabarti, R. and Jacobsen, S.B., The isotopic composition of magnesium in the inner Solar System. *Earth Planet. Sci. Lett.*, 293, 3–4, 349–358, 2010.
32. Dietze, M.C. *et al.*, Iterative near-term ecological forecasting: Needs, opportunities, and challenges. *Proc. Natl. Acad. Sci.*, 115, 7, 1424–1432, 2018.
33. Dickin, A.P., *Radiogenic isotope geology*, Cambridge University Press, Cambridge, England, 2018.
34. Cook, P.G. and Herczeg, A.L., *Environmental tracers in subsurface hydrology*, Springer Science & Business Media, Berlin/Heidelberg, Germany, 2012.
35. Waters, C.N. *et al.*, The Anthropocene is functionally and stratigraphically distinct from the Holocene. *Science (80-.)*, 351, 6269, aad2622, 2016.
36. Clark, I., *Groundwater geochemistry and isotopes*, CRC Press, Boca Raton, Florida, U.S.A., 2015.
37. Tan, C. *et al.*, Recent advances in ultrathin two-dimensional nanomaterials. *Chem. Rev.*, 117, 9, 6225–6331, 2017.
38. Huggett, R.J., *Climate, earth processes and earth history*, Springer Science & Business Media, Berlin/Heidelberg, Germany, 2012.
39. Armaroli, N. and Balzani, V., *Powering planet Earth: energy solutions for the future*, John Wiley & Sons, United States, 2012.

40. Cheng, H. *et al.*, Ice age terminations. *Science (80-.)*, 326, 5950, 248–252, 2009.
41. Kasting, J.F. and Howard, M.T., Atmospheric composition and climate on the early Earth. *Philos. Trans. R. Soc. B Biol. Sci.*, 361, 1474, 1733–1742, 2006.
42. Farquhar, G.D., Ehleringer, J.R., Hubick, K.T., Carbon isotope discrimination and photosynthesis. *Annu. Rev. Plant Biol.*, 40, 1, 503–537, 1989.
43. Hobson, K.A. and Wassenaar, L., II, *Tracking animal migration with stable isotopes*, Academic Press, Cambridge, Massachusetts, U.S.A., 2018.
44. McGlynn, G., *Using 13C-, 15N-, and 18O stable isotope analysis of human bone tissue to identify transhumance, high altitude habitation and reconstruct palaeodiet for the early medieval Alpine population at Volders, Austria*, lmu, Graz (Austria), 2007.
45. Roffet-Salque, M. *et al.*, From the inside out: Upscaling organic residue analyses of archaeological ceramics. *J. Archaeol. Sci. Rep.*, 16, 627–640, 2017.
46. McDermott, F., Palaeo-climate reconstruction from stable isotope variations in speleothems: a review. *Quat. Sci. Rev.*, 23, 7–8, 901–918, 2004.
47. Hoefs, J. and Hoefs, J., *Stable isotope geochemistry*, vol. 285, Springer, New York City, U.S.A., 2009.
48. Hoefs, J., Theoretical and experimental principles, in: *Stable isotope geochemistry*, pp. 1–46, Springer, New York City, U.S.A., 2015.
49. Vértes, A., Nagy, S., Klencsár, Z., Lovas, R.G., Rösch, F., *Handbook of Nuclear Chemistry: Vol. 1: Basics of Nuclear Science; Vol. 2: Elements and Isotopes: Formation, Transformation, Distribution; Vol. 3: Chemical Applications of Nuclear Reactions and Radiation; Vol. 4: Radiochemistry and Radiopharmaceutical Chemistry in Life Sciences; Vol. 5: Instrumentation, Separation Techniques, Environmental Issues; Vol. 6: Nuclear Energy Production and Safety Issues*, Springer Science & Business Media, New York City, U.S.A., 2010.
50. Lane, N., *The vital question: energy, evolution, and the origins of complex life*, WW Norton & Company, New York, U.S.A., 2015.
51. Reilly, C., *The nutritional trace metals*, John Wiley & Sons, New Jersey, U.S.A., 2008.
52. Sharma, P.D., *Environmental Biology & Toxicology*, Rastogi Publications, Shivaji Road Meerut - 250 002, U.P. India, 2014.
53. Atwood, D.A., *The rare earth elements: fundamentals and applications*, John Wiley & Sons, New Jersey, U.S.A., 2013.
54. Sverdrup, H.U. and Ragnarsdóttir, K.V., Section 2. Classification of natural resources. *Geochem. Perspect.*, 3, 2, 172–192, 2014.
55. Prashanth, L., Kattapagari, K.K., Chitturi, R.T., Baddam, V.R.R., Prasad, L.K., A review on role of essential trace elements in health and disease. *J. Dr. NTR Univ. Health Sci.*, 4, 2, 75, 2015.
56. Badmaev, V., Prakash, S., Majeed, M., Vanadium: a review of its potential role in the fight against diabetes. *J. Altern. Complement. Med.*, 5, 3, 273–291, 1999.
57. N.C. Loizou, Health supplement. Google Patents, 30-Apr-2009.
58. Jiang, P. *et al.*, Effect of vanadyl rosiglitazone, a new insulin-mimetic vanadium complexes, on glucose homeostasis of diabetic mice. *Appl. Biochem. Biotechnol.*, 180, 5, 841–851, 2016.

59. De Cremer, K., 2.21 Speciation of Vanadium, in: *Handb. Elem. Speciat. II-Species Environ. Food, Med. Occup. Heal*, 464, 2005.
60. Okoro, H.K., Fatoki, O.S., Adekola, F.A., Ximba, B.J., Snyman, R.G., A review of sequential extraction procedures for heavy metals speciation in soil and sediments. Open Access Scientific Reports, 1, 3, 1–9, 2012.
61. Cámara-Martos, F. and Moreno-Rojas, R., Cobalt: Properties and Determination. *Encyclopedia of Food and Health*, 166–171, 2016.
62. Fulekar, M.H., *Environmental biotechnology*, CRC Press, Boca Raton, Florida, U.S.A., 2010.
63. Alloway, B.J., *Heavy metals in soils: trace metals and metalloids in soils and their bioavailability*, vol. 22, Springer Science & Business Media, Berlin/Heidelberg, Germany, 2012.
64. Grandjean, P. and Landrigan, P.J., Developmental neurotoxicity of industrial chemicals. *Lancet*, 368, 9553, 2167–2178, 2006.
65. Dongarrà, G., Tamburo, E., Varrica, D., Dust, Metals and Metalloids in the Environment: From Air to Hair, in: *Medical Geochemistry*, pp. 127–148, Springer, Berlin/Heidelberg, Germany, 2013.
66. Durant, A.J., Bonadonna, C., Horwell, C.J., Atmospheric and environmental impacts of volcanic particulates. *Elements*, 6, 235–240, 2010.
67. Nordberg, M. and Cherian, M.G., I. An Introduction to Biological Responses of Certain Elements. *Med. Geol.*, In: *Essentials of Medical Geology Impacts of the Natural Environment on Public Health*, Elsevier, Cambridge, Massachusetts, U.S.A., 179, 2005.
68. Khudhair, A.Y., Determination of trace metals in hair samples of Iraqis people living in suburban and urban in Basrah governate, southern of Iraq. *J. Thi-Qar Sci.*, 4, 3, 58–67, 2014.
69. Goldey, E.S., Kehn, L.S., Lau, C., Rehnberg, G.L., Crofton, K.M., Developmental exposure to polychlorinated biphenyls (Aroclor 1254) reduces circulating thyroid hormone concentrations and causes hearing deficits in rats. *Toxicol. Appl. Pharmacol.*, 135, 1, 77–88, 1995.
70. Louis, L.M. *et al.*, A prospective study of cancer risk among Agricultural Health Study farm spouses associated with personal use of organochlorine insecticides. *Environ. Health.*, 16, 1, 95, 2017.
71. Kelton, J.A., *Herbicides-current research and case studies in use*, 2013.
72. Lushchak, V., II, Matviishyn, T.M., Husak, V.V., Storey, J.M., Storey, K.B., Pesticide toxicity: a mechanistic approach. *EXCLI J.*, 17, 1101, 2018.
73. Leyla, B., Mostafa, A., Sasan, A., Mehdi, S., Tao, L., Oksana, L., Effects of growth under different light spectra on the subsequent high light tolerance in rose plantsm. *AoB Plants*, 12, 10, 5, ply052, 2018.
74. Dayan, F.E. and Duke, S.O., Natural compounds as next-generation herbicides. *Plant Physiol.*, 166, 3, 1090–1105, 2014.
75. Gomes, M.P. *et al.*, Glyphosate-dependent inhibition of photosynthesis in willow. *Front. Plant Sci.*, 8, 207, 2017.

76. Kjellstrom, T., Lodh, M., McMichael, T., Ranmuthugala, G., Shrestha, R., Kingsland, S., Air and water pollution: burden and strategies for control, in: *Disease Control Priorities in Developing Countries*, 2nd edition, The International Bank for Reconstruction and Development/The World Bank, Washington, DC, USA, 2006.
77. Adjei, J.K., Essumang, D.K., Twumasi, E., Nyame, E., Muah, I., Levels and risk assessment of residual phthalates, polycyclic aromatic hydrocarbons and semi-volatile chlorinated organic compounds in toilet tissue papers. *Toxicol. Rep.*, 6, 1263–1272, 2019.
78. Anku, W.W., Mamo, M.A., Govender, P.P., Phenolic compounds in water: sources, reactivity, toxicity and treatment methods, in: *Phenolic Compd. sources. Importance Appl. First Ed.*, pp. 419–443, InTech, London, UK., 2017.
79. Faroon, O. and Ruiz, P., Polychlorinated biphenyls: New evidence from the last decade. *Toxicol. Ind. Health*, 32, 11, 1825–1847, 2016.
80. Masindi, V. and Muedi, K.L., Environmental contamination by heavy metals, in: *Heavy Met*, pp. 115–133, IntechOpen, Aglan, Fr, 2018.
81. Speight, J., Sources and Types of Inorganic Pollutants. *Environ. Inorg. Chem. Eng. Sect.*, 5, 2, 231–282, 2017.
82. Richard, A.M., Diaz, J.H., Kaye, A.D., Reexamining the risks of drinking-water nitrates on public health. *Ochsner J.*, 14, 3, 392–398, 2014.
83. Kulawik, P. and Kumar Tiwari, B., Recent advancements in the application of non-thermal plasma technology for the seafood industry. *Crit. Rev. Food Sci. Nutr.*, 59, 19, 3199–3210, 2019.
84. Park, J.H., Bolan, N.S., Chung, J.W., Naidu, R., Megharaj, M., Environmental monitoring of the role of phosphate compounds in enhancing immobilization and reducing bioavailability of lead in contaminated soils. *J. Environ. Monit.*, 13, 8, 2234–2242, 2011.
85. Khanna, R. and Gupta, S., Agrochemicals as a potential cause of ground water pollution: a review. *Int. J. Chem. Stud.*, 6, 985–990, 2018.
86. Ag, M. and Am, E.S., Current situation of water pollution and its effect on aquatic life in Egypt. *Egypt. J. Occup. Med.*, 37, 1, 95–115, 2013.
87. Gheorghe, I.F. and Ion, B., The effects of air pollutants on vegetation and the role of vegetation in reducing atmospheric pollution, in: *impact air Pollut. Heal. Econ. Environ. Agric. sources*, pp. 241–280, 2011.
88. Cichowicz, R., Wielgosiński, G., Fetter, W., Dispersion of atmospheric air pollution in summer and winter season. *Environ. Monit. Assess.*, 189, 12, 605, 2017.
89. Zhang, Z., Wang, J., Lu, W., Exposure to nitrogen dioxide and chronic obstructive pulmonary disease (COPD) in adults: a systematic review and meta-analysis. *Environ. Sci. Pollut. Res.*, 25, 15, 15133–15145, 2018.
90. Delworth, T.L., Zeng, F., Vecchi, G.A., Yang, X., Zhang, L., Zhang, R., The North Atlantic Oscillation as a driver of rapid climate change in the Northern Hemisphere. *Nat. Geosci.*, 9, 7, 509, 2016.

9

Inorganic Geochemistry

Sathasivam Pratheep Kumar[1], Triveni Rajashekhar Mandlimath[1] and M. Ramesh[2]*

[1]*Department of Chemistry, KPR Institute of Engineering and Technology, Coimbatore, Tamil Nadu, India*
[2]*Department of Mechanical Engineering, KIT-Kalaignarkarunanidhi Institute of Technology, Coimbatore, Tamil Nadu, India*

Abstract
Naturally occurring earth minerals have potential applications in various fields due to their interesting properties. Study of inorganic geochemistry is much needed for the geological scientists and researchers to identify the elements and compositions of natural minerals, which exist in the earth. Hence, in this chapter, the importance of inorganic geochemistry, existence of various periodic elements, geological minerals in the earth, and their structural difference are discussed. Characterization of the elements and minerals in geology is significant for considering them in practical applications. The sophisticated analytical methods such as powder X-ray diffraction, scanning electron microscopy, electron probe micro-analysis, etc., for the characterization of inorganic or geological minerals are presented. Further, the importance of each technique, sample preparation, and their detection limits are discussed.

Keywords: Inorganic geochemistry, geological minerals, clay minerals, scanning electron microscopy, electron probe micro-analysis

9.1 Introduction

Inorganic geochemistry is a science which provides useful insights into many lithological and geo-chemical works and has important applications in igneous, sedimentary, and metamorphic petrology [1]. Geochemistry is

**Corresponding author*: mramesh97@gmail.com

Inamuddin, Mohd Imran Ahamed, Rajender Boddula and Tariq Altalhi (eds.) Geochemistry: Concepts and Applications, (149–160) © 2021 Scrivener Publishing LLC

the study of chemical processes, which are involved in naturally existing earth materials like rock and analysis of their compositions and structure. Majority of the geological materials are inorganic minerals. Geological minerals are considered as the building blocks of rock. Rocks are formed by the combination of minerals. There are several thousands of natural minerals available in the earth [2]. Earth is mainly consisting of core (inner and outer), mantle, and crust. Its upper portion (crust) is divided into three categories of rocks such as i) igneous rock (basalt, granite, gabbro, and rhyolite) formed by the cooling of molten silicates (magma) inside the earth, ii) sedimentary rock (sandstone and limestone) formed by the weathering and eroded components of other rocks, and iii) metamorphic rocks (gneiss, schist, and slate) formed by changing the pressure and temperature inside the earth [2, 3]. Among the huge number of naturally existing minerals, the main rock forming minerals are quartz, feldspars, amphiboles, pyroxenes, olivines, and calcite. All these minerals are in the form of silicates other than calcite. Majority of the minerals are composed of Al, Si, and O elements since they are abundant in the earth crust. Geochemical analysis of naturally occurring minerals and periodic elements is important as they are refractory materials and their concentration varies from parts per million (ppm) to parts per billion (ppb) [2–4]. In order to identify these elements or minerals accurately, most precise instrumentation techniques are required for geological research. Hence, in this chapter, the existence of geological inorganic minerals and their applications, periodic elements, and their chemical analysis by various instrumentation techniques are presented.

9.2 Elements and the Earth

The geochemical composition of earth materials is essential to support our understanding of planetary dynamics, the exploitation of resources, and anthropogenic impacts [5]. Among all the periodic elements, 94 are naturally occurring in the earth. However, Fe, O, Si, and Mg alone make more than 90% of the earth's mass. The most abundant elements in the earth's crust are O, Si, Al, Fe, Ca, Na, Mg, and P. Table 9.1 summarizes the elements available in the earth and their concentrations in ppm or wt. % or ppb [3].

9.2.1 Iron

Iron (Fe) is the most abundant element in the earth. The major part of Fe is found in the inner core of the earth. The outer core and mantle contains

Table 9.1 Elemental composition of the earth [3].

Element	Concentration (wt. % or ppm or ppb)	Element	Concentration (wt. % or ppm or ppb)	Element	Concentration (wt. % or ppm or ppb)
K	162	Y	2.9	Pr	0.171
Ca	1.71	Zr	7.1	Nd	0.844
Sc	10.93	Nb	0.444	Sm	0.274
Ti	813	Mo	1660	Eu	0.104
V	94.4	Ru	1310	Gd	0.367
Cr	3720	Rh	242	Tb	0.067
Mn	1680	Pd	1015	Dy	0.455
Fe	30.3	Ag	54.4	Ho	0.101
Co	838	Cd	76	Er	0.296
Ni	17,600	In	6.8	Tm	0.046
Cu	60.9	Sn	251.1	Yb	0.297
Zn	37.1	Sb	46.2	Lu	0.046
Ga	2.7	Te	285.7	Hf	0.191
Ge	7.24	I	6.8	Ta	0.025
As	1.669	Xe	–	W	0.173
Se	2.65	Cs	0.014	Re	75.3
Br	3.57	Ba	4.445	Os	900
Kr	–	La	0.437	Ir	455
Rb	0.405	Ce	1.131	Pt	1866
Sr	13.4				

Fe in the form of alloy and silicates. All together, the earth's mass consisting of ~35 % Fe.

9.2.2 Oxygen

The higher percentage of oxygen is found in the crust. It is also a major component in the earth's atmosphere which is approximately 21%. It is highly reactive and readily forms oxides with the metal ions. It accounts around 30 % of the mass of the entire earth.

9.2.3 Silicon

Silicon is found in the form of compounds in mantle and crust. It forms compounds with Fe, Mg, and O. Most of the compounds are available as natural minerals such as quartz, feldspar, amphibole, etc. It exists around 15 % in the earth's mass.

9.2.4 Magnesium

Magnesium (Mg) is available in the earth's crust as minerals and clay minerals. Few examples of Mg compounds are dolomite, talc, olivine, etc. In total, 13% of Mg is present in the earth's mass.

9.3 Geological Minerals

In this section, the compositions, structural arrangements, and potential applications of major rock forming minerals are described. The crystal structures of these minerals are mainly composed of SiO_4 tetrahedral and their linkage as shown in Figure 9.1.

9.3.1 Quartz

Quartz (SiO_2) is the most abundant mineral in the earth's igneous, sedimentary, and metamorphic rocks. It has a large number of applications in various industries with the lead in glass and electronic industry.

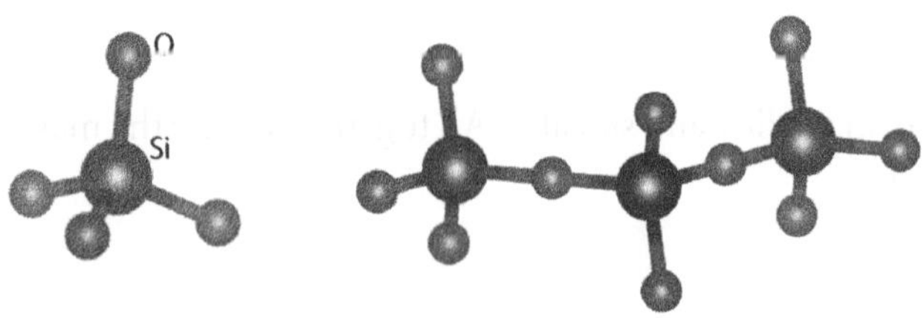

Figure 9.1 SiO_4 tetrahedra and its network.

The structure consists of SiO_4 tetrahedral framework, which forms by the sharing of O with Si atoms.

9.3.2 Feldspar

The major composition of feldspar is sodium, potassium, and calcium alumino silicate (AB_4O_8; A = Na, K, Ca; B-Al, Si) [6]. The combination of these minerals makes a large part of the earth crust. The three-dimensional framework structures are composed of SiO_4 tetrahedra. Feldspars have a wide range of applications in ceramic and glass industry.

9.3.3 Amphibole

Amphibole ($Ca_2(Fe, Mg)_4Al(Si_2Al)O_{22}(OH, F)_2$) exists mainly in the earth's igneous and metamorphic rocks. The structure formed by the double chains of SiO_4 tetrahedra. The major constituents of amphiboles are Ma, Fe, Mn, and alumino-silicates [6]. Since this mineral is harder, it is widely being used for construction and bedding purposes.

9.3.4 Pyroxene

Pyroxene ($(Ca, Na)(Mg, Fe, Al, Ti)(Si, Al)_2O_6$) minerals are the ionosilicates similar to the amphiboles. The major difference between amphiboles and pyroxenes is the structural arrangement [6]. Pyroxene structure is formed by the single chain of SiO_4 tetrahedra, whereas amphiboles are double chain silicates.

9.3.5 Olivine

Olivine $(Mg/Fe)_2SiO_4$ is a type of ortho-silicate forms based on the composition of forsterite (Mg_2SiO_4) and fayalite (Fe_2SiO_4). The layered structure consists of Fe/Mg octahedra and SiO_4 tetrahedra [6]. Olivines are being used as gemstone and refractory mineral in steel industry.

9.3.6 Clay Minerals

Clay minerals are the layered silicate structures. The structural type of clay minerals varies as 1:1 and 2:1 depending upon the crystallographic arrangement of tetrahedral and octahedral sheets. Figure 9.2 illustrates the polyhedral structural representation of 1:1 and 2:1 clay minerals [7].

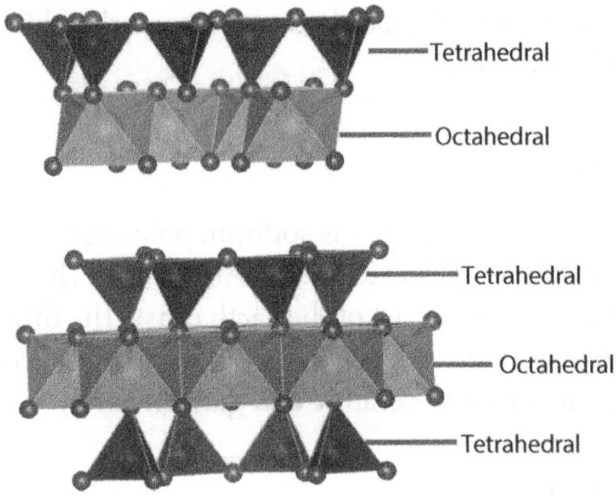

Figure 9.2 Polyhedral projection of 1:1 and 2:1 clay minerals.

The negative charge which creates due to the isomorphous substitutions in the tetrahedral and octahedral sheets is balanced by the interlayer cations. Rapid heating of these clay minerals can lead to several times volume expansion due to the sudden release of interlayer water molecules. Clay minerals show important properties such as low density, chemical stability, and ion exchange capability. They have significant role in various industries. They have been used as primary fillers to various polymer-based intumescent materials to enhance the flame retardancy of polymers, catalyst, and refractory materials.

9.3.7 Kaolinite

Kaolinites are classified as 1:1 clay mineral as they are composed of one tetrahedral sheet linked to an octahedral sheet. Since kaolin is a non-expandable mineral, it cannot absorb the water molecules in between the tetrahedral and octahedral sheets [7].

9.3.8 Bentonite, Montmorillonite, Vermiculite, and Biotite

These minerals are naturally exist in the form of hydrated magnesium aluminum silicates and classified as 2:1 clay minerals. The structures are composed of two tetrahedral sheets linked with one octahedral sheet [7]. The octahedral sheet is sandwiched between two tetrahedral sheets and forms an expandable layered structure. These minerals have wide range of applications in the field of chemical, agricultural industries, building materials, and environmental protection because of its excellent properties such as light weight, abundant supply, cheap cost, low density, thermal stability, chemical stability, good cation exchanger, and adsorption capability.

9.4 Characterization Techniques

Characterization of elements and minerals involves analysis of the phase, composition, structure, morphology, etc. The identification of geological elements and minerals is possible by chemical and analytical methods [8]. The compositional analysis of alkaline earth metals and light elements such as H, C, N, S, and O are challenging as they can form different compounds at elevated temperatures; for example, metal carbonates, silicates, and oxides. Similarly, analysis of transition elements is another important task since they form sulfides and release toxic chemicals to the environment and the soil. Additionally, transition element–based compounds are being used in many industries like dye, textile, and chemical processing. In addition to the analysis of light elements and alkaline earth metals, identification of rare earth elements is significant as they are available trace level in the earth. The analytical techniques such as powder X-ray diffraction (XRD), X-ray fluorescence spectra (XRF), X-ray photoelectron spectra (XPS), electron probe micro-analysis (EPMA), inductively coupled plasma (ICP) spectroscopy, and electron microscopic techniques are the major instrumental methods for the determination of inorganic elements and minerals. The following section covers the geological sample collection, sampling processes, analysis procedure, and the importance of each technique.

9.4.1 Powder X-Ray Diffraction

XRD is one of the important and primary techniques to identify the phases present in the mineral. The geological mineral oxides, carbonates, silicates, phosphates, etc., can be identified by this method. It is a non-destructive technique, requires very minimal quantity of the sample, and does not need any specific sample preparation. From the XRD patterns, unit cell parameters, cell volume, and other structural descriptions such as bond length, bond angle, atomic positions, and particle sizes can be calculated. The International Centre for Diffraction Data (ICDD) is much useful for the determination of new minerals available in the rock. By identifying the mineral phases present in the earth materials, composition of each phase can also be determined by this method [9]. The amorphous materials like glasses and liquids cannot be identified by this technique. Single crystal XRD is used for the structural analysis of crystals.

9.4.2 X-Ray Fluorescence Spectra

Multi elemental analysis is possible from the XRF. Though the solid, powder, and liquid samples can be analyzed by this method, solid samples with flat surface are preferred to get the accurate results. The detectable limit of elements by this technique is μg/g and is not suitable beyond this level. Geological mineral powders are suitable for analysis when they compressed into pellet form with smooth flat surface. For the instrumental calibration oxide or carbonate standards are being used. Sample holders may vary based on the sample natures [10]. XRF is widely used in research institutes and chemical industries for the determination of multi elements apart from the light elements such as hydrogen, helium and lithium.

9.4.3 X-Ray Photoelectron Spectra

Electron spectroscopy for chemical analysis or XPS is a powerful technique for the quantitative determination of chemical composition and chemical state on the surface of a material. In this method, up to 10-nm depth of the solid surface can be analyzed more precisely. This technique involves the interaction between the soft X-ray and the material's surface, which emits the photoelectrons from the outer orbitals. The emitted energy can be compared with the standard energy values to find the chemical state of the elements. This technique is widely being used in geo-chemical analysis to detect the trace amount of elements and mineral compositions [11]. Textile industries, chemical industries, and research institutes are the major users of XPS. Similar to XRF all elements can be identified by this method except hydrogen and helium. It is highly surface sensitive technique and requires ultra-high vacuum.

9.4.4 Electron Probe Micro-Analysis

EPMA, popularly known as electron probe in the geological research, is the foremost analytical technique for the quantitative and qualitative analysis of elements and minerals. However, this technique is mostly being used by mineralogical and geological scientists for quantitative analysis of elements present in the various types of rocks. The point, linear, and area analysis types of EPMA are useful to detect the elements in certain points of the material surface, linear elemental change, and elements distributions, respectively. Sample preparation for the analysis involves mixing of polymer-based resins to make well-polished solid surface. Electron microprobe combines with the microscopic techniques to yield the micrographs, which are useful for the identification of mineral grains, textures, and compositions. In particular, the back scattered electron images are most commonly used for the efficient detection of different compositions [12, 13]. The combination of EPMA and powder XRD is useful for the determination of the mineral phases and compositions. The detection limit of elements by this method is 100 mg/kg.

9.4.5 Inductively Coupled Plasma Spectrometry

Different types of ICP spectrometric techniques such as ICP–optical emission spectroscopy (ICP-OES), ICP–atomic emission spectroscopy (ICP-AES), and ICP–mass spectroscopy (ICP-MS) are familiar for the elemental analysis of liquid samples. ICP spectroscopy is one of the important analytical techniques in the atomic spectroscopy to detect the trace elements down to ppb levels in the geological minerals [14, 15]. Here, plasma (ionized gas) is used as excitation source for atomization. The liquid samples are sprayed through the nebulizer and converted into aerosol. The aerosol gets atomized and the atoms emit energy in particular wavelength, which can be measured and compared with the standard values. The temperature range of plasma is from 600 to 6000 K. Solid oxide or carbonate samples need some specific sample preparation with the solvents like nitric acid, alcohol, etc. Unlike other absorption and emission techniques, ICP spectroscopy can detect all the elements except inert gases, environmental gases and some non-metals like C, N, H, and O. Simultaneous detection of around 30 elements is possible by this technique.

9.4.6 Fourier Transform Infrared Spectroscopy

Fourier transform infrared (FTIR) spectroscopy is an analytical tool being used in geology for analyzing the geological samples. The infrared source is used to excite the molecules present in the sample to the higher energy state due to the absorption of light. This led to the vibration of molecules and results a characteristic stretching and bending absorption bands, which can be compared with the standard functional group peak values for the determination of the structure. Geochemists mainly use this technique to identify the functional groups of geological minerals such as OH, CO, Si-O, and P-O and other metal oxide peaks [16]. Since majority of the geological minerals are found in the form of silicates, FTIR is much useful method to detect the silicate group. The characteristic vibrational modes of SiO_4 tetrahedra appear in the region 450–1220 cm^{-1}. The carbonyl and hydroxyl groups are fall at ~1,500 cm^{-1} and ~3,500 cm^{-1}, respectively. The metal oxide bands are generally fall in the lower wavenumber region (400–500 cm^{-1}). Most of the cases, the mid IR (400–4,000 cm^{-1}) is used for the analysis of geological samples.

9.4.7 Scanning Electron Microscopy Analysis

In geochemical analysis, scanning electron microscopy (SEM) is used to observe the surface morphology of minerals. A focused beam of electrons from electron gun probes the surface structure of the samples and yield various parameters such as microstructures, composition, and structural information. SEM combined with energy dispersive X-ray spectroscopy (EDX), back scattered detector (BSD) and electron back scattering diffraction (EBSD) and provides elemental concentration, BSE image and structural features respectively. The high resolution micrographs can be obtained from field emission SEM (FE-SEM). From the micrographs, porosity of a material, shape and size of the particles and the defects in the material can be obtained. All solid and powder samples can be analyzed by this technique. Sputtering of conducting metal is used for semiconductors and insulators samples. The low and ultra-vacuum modes of SEM are much useful to analyze the geological samples without sputtering [17]. More recently, concentration of the elements are analyzed by electron mapping method to get accurate results. For internal structure analysis transmission electron microscopy (TEM) is used with the high energy electrons.

9.4.8 Energy Dispersive X-Ray Analysis

EDX or EDAX is another important quantitative analysis for the detection of elemental percentage in geological minerals. EDX combined with microscopic techniques like SEM and TEM and yield the element concentrations on various locations of the surface of the samples [18]. In principle, when the beam of electrons hits the sample, the inner electron is kicked out and the vacancy is filled by the outer electron. During this electron transfer, the released X-rays are measured in terms of energy for each metal ion. Almost all the metal ions can be analyzed by this method except the light elements such as H, N, etc. EDX requires very small quantity of the samples and does not need any specific sample preparation.

9.5 Conclusion

The importance of inorganic geochemistry, the rock forming minerals, and their compositions have been discussed in this chapter. The earth consists of number of elements in ppm or ppb levels. Al, Si, and O are forming majority of the minerals in the earth crust. The structure of minerals is mainly composed by SiO_4 tetrahedra and its network with the metal octahedra. Clay minerals with 1:1 and 2:1 sheet like structures and their applications have been explained with the polyhedral structural model. The characterization of minerals, elemental analysis, and the determination of compositions have been elaborated with the powder XRD, SEM, FTIR, XRF, XPS, EPMA, and ICP spectroscopic techniques.

References

1. Eric, B.E., Emile, E., Konfor, N.I., Fralick, P., Betrant, B.S., Ntoboh., T.C., Inorganic geochemistry and petroleum source evaluation of organic black shale in the Mamfe Basin (West Africa). *Solid Earth Sci.*, 4, 166–177, 2019.
2. White, W.M., *Geochemistry*, Wiley Blackwell, Chichester, 2013.
3. Trueman, C.N., Rodgers, K.J., McLellan, I.S., Hursthouse, A.S., Geochemistry/Inorganic, in: *Encyclopedia of Analytical Science*, pp. 171–181, Elsevier Ltd., UK 2005.
4. Haldar, S.K., Geology and Geochemistry, in: *Platinum-Nickel-Chromium Deposits: Geology, Exploration, and Reserve Base*, pp. 37–61, 2017.
5. Trueman, C.N., Rodgers, K.J., McLellan, I.S., Hursthouse, A.S., Geochemistry inorganic. reference module in chemistry, Molecular Sciences and Chemical Engineering, in: *Encyclopedia of Analytical Science*, Third Edition, pp. 271–282, 2019, https://doi.org/10.1016/B978-0-12-409547-2.14362-8.
6. Earle, S., *Physical Geology*, 2nd Edition, BCcampus, Victoria, B.C. 2019, https://opentextbc.ca/physicalgeolog.
7. Dill, H.G., A geological and mineralogical review of clay mineral deposits and phyllosilicate ore guides in Central Europe – A function of geodynamics and climate change. *Ore Geol. Rev.*, 119, 103304, 2020.
8. Vutchkov, M., Lalor, G., Macko, S., Inorganic and Organic Geochemistry Techniques, in: *Essentials of Medical Geology*, pp. 689–716, 2012.
9. Lavina, B., Dera, P., Downs, R.T., Modern X-ray Diffraction Methods in Mineralogy and Geosciences. *Rev. Mineral. Geochem.*, 78, 1–31, 2014.
10. Nakayama, K. and Nakamura, T., X-Ray Fluorescence Spectroscopy for Geochemistry, in: *Treatise on Geochemistry*, pp. 181–194, 2014.
11. Salama, W., Mourtada, E.A., Gaupp, R., Spectroscopic characterization of iron ores formed in different geological environments using FTIR, XPS,

Mössbauer spectroscopy and thermoanalyses. *Spectrochim. Acta Part A: Mol. Biomol. Spectrosc.*, 136, 1816–1826, 2015.
12. He, Q. and Shi, C., Application of electron probe micro analyzer (EPMA) to depositional, environment identification of sedimentary rock. *Adv. Mat. Res.*, 881, 1795–1798, 2014.
13. McGeel, J.J. and Keil, K., Application of electron probe microanalysis to the study of geological and planetary materials. *Microsc. Microanal.*, 7, 200–210, 2001.
14. Navarro, M.S., Ulbrich, H.H.G.J., Andrade, S., Janasi, V.A., A daptation of ICP–OES routine determination techniques for the analysis of rare earth elements by chromatographic separation in geologic materials: tests with reference materials and granitic rocks. *J. Alloys Compd.*, 344, 40–45, 2002.
15. Ketterer, M.E., Geology and mineralogy applications of atomic spectroscopy, in: *Encyclopedia of Spectroscopy and Spectrometry*, pp. 25–29, 2017.
16. Chen, Y., Zou, C., Mastalerz, M., Hu, S., Gasaway, C., Tao, X., Applications of micro-fourier transform infrared spectroscopy (FTIR) in the geological sciences-A review. *Int. J. Mol. Sci.*, 16, 30223–30250.
17. Chen, L., Xu, J., Chen, J., Applications of scanning electron microscopy in earth sciences. *Sci. China Earth Sci.*, 58, 1768–1778, 2015, https://doi.org/10.1007/s11430-015-5172-9.
18. Olabanji, S.O., Haque, A.M., II, Zandolin, S., Ajayi, R.T., Buoso, M.C., Ceccato, D., Cherubini, R., Zafiropoulos, D., Moschini, G., Applications of PIXE, PIGE, SEM and EDAX techniques to the study of geological samples. *Nucl. Instrum. Methods Phys. Res. Sec. B: Beam Interact. Mater. At.*, 109-110, 262–265, 1996.

10

Introduction and Scope of Geochemistry

Triveni Rajashekhar Mandlimath[1], Sathasivam Pratheep Kumar[1] and M. Ramesh[2]*

[1]Department of Chemistry, KPR Institute of Engineering and Technology, Coimbatore, Tamil Nadu, India
[2]Department of Mechanical Engineering, KIT-Kalaignarkarunanidhi Institute of Technology, Coimbatore, Tamil Nadu, India

Abstract
Geochemistry is the broad captivating area of science which utilizes the principles of chemistry as a tool to analyze the elemental composition and chemical reactions both on earth and the planets. The geochemical knowledge is highly required to solve the numerous environmental issues. Most of the elements are naturally exist in the earth in the form of rocks, minerals, and ores. Hence, it is necessary to study the physico-chemical properties of periodic elements and distribution of geochemical classification elements. Thus, this chapter describes the periodic elements and their physic-chemical properties and chemical composition of various earth's layers and spheres. Several sub-disciplines and scope of geochemistry are discussed.

***Keywords*:** Geochemistry, periodic properties, layers, atmosphere, sub-disciplines

10.1 Introduction

The word geochemistry was first used in the year 1838 by the German-Swiss chemist Christian Friedrich Schonbein [1]. In general, geochemistry term is a combination of geology and chemistry which means the study of earth's chemical composition. American geochemist Frank W. Clarke who was the chief chemist of United States geological survey contributed many articles based on the composition of rocks, minerals, and ores, and

**Corresponding author*: mramesh97@gmail.com

Inamuddin, Mohd Imran Ahamed, Rajender Boddula and Tariq Altalhi (eds.) Geochemistry: Concepts and Applications, (161–178) © 2021 Scrivener Publishing LLC

also the chemical composition of earth's crust [2]. Since then, he is recognized as a founder of geochemistry. It was majorly known for studying the composition of minerals and rock still 1950s [3]. Modern geochemistry has emerged after Victor Goldschmidt's (known as father of modern geochemistry) remarkable contribution on the investigation of the abundance of elements in meteorites and cosmos. His research on the distribution of elements, the role of ionic, atomic substitution, and coordination in crystal lattice has left new directions to geochemists [2]. The USSR geochemists revolutionized the analytical techniques to determine the earth's components with high precision [2]. Modern geochemistry often deals with the systematic study of distribution of elements, ratio of isotopes in natural materials, and varieties of chemical processes in the earth and other planets both at present and in the past.

10.1.1 Periodic Table and Electronic Configuration

10.1.1.1 Periodic Table

In the year 1871, Russian chemist Dmitri Mendeleev arranged the elements in the periodic table based on the atomic number in an ascending order. The elemental properties are explained on the basis of their atomic weights. Figure 10.1 represents the periodic table of the elements. In the periodic table, the elements organized in the same group exhibit similar chemical properties. Atomic number of an element indicates either the number of electrons or protons. The atomic weight corresponds to the number of protons and neutrons present in the nucleus. In an atom, nucleus at the center is composed of neutrons and protons. Electrons revolve around the nucleus. The atoms with the similar atomic number but different mass number are called as isotopes.

For example, ^{3}He and ^{4}He are the two stable isotopes of ^{2}He. Both the isotopes own two protons, whereas the number of neutrons in ^{3}He and ^{4}He is 1 and 2, respectively. In the periodic table, the elements are named as s, p, d, and f block elements based on the type of atomic orbitals being filled with electrons. The s block consists of alkali and alkaline earth metals; p block includes metals, non-metals, halogens, and inert gases; d block comprises transition metals and noble metals; and f block is with rare earth elements called as lanthanides and actinides. The physical properties such as atomic radii, ionic radii, electro-negativity, and first ionization energy vary periodically in the periodic table.

Key: atomic number / Symbol / name / conventional atomic weight / standard atomic weight

1	2	3	4	5	6	7	8	9	10	11	12	13	14	15	16	17	18
1 H hydrogen 1.008 [1.0078, 1.0082]																	2 He helium 4.0026
3 Li lithium 6.94 [6.938, 6.997]	4 Be beryllium 9.0122											5 B boron 10.81 [10.806, 10.821]	6 C carbon 12.011 [12.009, 12.012]	7 N nitrogen 14.007 [14.006, 14.008]	8 O oxygen 15.999 [15.999, 16.000]	9 F fluorine 18.998	10 Ne neon 20.180
11 Na sodium 22.990	12 Mg magnesium 24.305 [24.304, 24.307]											13 Al aluminium 26.982	14 Si silicon 28.085 [28.084, 28.086]	15 P phosphorus 30.974	16 S sulfur 32.06 [32.059, 32.076]	17 Cl chlorine 35.45 [35.446, 35.457]	18 Ar argon 39.95 [39.792, 39.963]
19 K potassium 39.098	20 Ca calcium 40.078(4)	21 Sc scandium 44.956	22 Ti titanium 47.867	23 V vanadium 50.942	24 Cr chromium 51.996	25 Mn manganese 54.938	26 Fe iron 55.845(2)	27 Co cobalt 58.933	28 Ni nickel 58.693	29 Cu copper 63.546(3)	30 Zn zinc 65.38(2)	31 Ga gallium 69.723	32 Ge germanium 72.630(8)	33 As arsenic 74.922	34 Se selenium 78.971(8)	35 Br bromine 79.904 [79.901, 79.907]	36 Kr krypton 83.798(2)
37 Rb rubidium 85.468	38 Sr strontium 87.62	39 Y yttrium 88.906	40 Zr zirconium 91.224(2)	41 Nb niobium 92.906	42 Mo molybdenum 95.95	43 Tc technetium	44 Ru ruthenium 101.07(2)	45 Rh rhodium 102.91	46 Pd palladium 106.42	47 Ag silver 107.87	48 Cd cadmium 112.41	49 In indium 114.82	50 Sn tin 118.71	51 Sb antimony 121.76	52 Te tellurium 127.60(3)	53 I iodine 126.90	54 Xe xenon 131.29
55 Cs caesium 132.91	56 Ba barium 137.33	57-71 lanthanoids	72 Hf hafnium 178.49(2)	73 Ta tantalum 180.95	74 W tungsten 183.84	75 Re rhenium 186.21	76 Os osmium 190.23(3)	77 Ir iridium 192.22	78 Pt platinum 195.08	79 Au gold 196.97	80 Hg mercury 200.59	81 Tl thallium 204.38 [204.38, 204.39]	82 Pb lead 207.2	83 Bi bismuth 208.98	84 Po polonium	85 At astatine	86 Rn radon
87 Fr francium	88 Ra radium	89-103 actinoids	104 Rf rutherfordium	105 Db dubnium	106 Sg seaborgium	107 Bh bohrium	108 Hs hassium	109 Mt meitnerium	110 Ds darmstadtium	111 Rg roentgenium	112 Cn copernicium	113 Nh nihonium	114 Fl flerovium	115 Mc moscovium	116 Lv livermorium	117 Ts tennessine	118 Og oganesson

57 La lanthanum 138.91	58 Ce cerium 140.12	59 Pr praseodymium 140.91	60 Nd neodymium 144.24	61 Pm promethium	62 Sm samarium 150.36(2)	63 Eu europium 151.96	64 Gd gadolinium 157.25(3)	65 Tb terbium 158.93	66 Dy dysprosium 162.50	67 Ho holmium 164.93	68 Er erbium 167.26	69 Tm thulium 168.93	70 Yb ytterbium 173.05	71 Lu lutetium 174.97
89 Ac actinium	90 Th thorium 232.04	91 Pa protactinium 231.04	92 U uranium 238.03	93 Np neptunium	94 Pu plutonium	95 Am americium	96 Cm curium	97 Bk berkelium	98 Cf californium	99 Es einsteinium	100 Fm fermium	101 Md mendelevium	102 No nobelium	103 Lr lawrencium

Figure 10.1 Periodic table.

10.1.1.2 Electronic Configuration

The electronic configuration is the arrangement of electrons in atomic orbitals. The periodic table represents the principal quantum number "n" value for the valence or outermost shell. The first period consists of two elements such as hydrogen ($1s^1$) and helium ($1s^2$) in which the electron enter the first K shell. The second period (n = 2) comprised of eight elements, begins with lithium which has the electronic configuration $1s^22s^1$. The electronic configuration of beryllium is $1s^2\ 2s^2$. From boron, electrons fill the 2p orbital. This period ends with neon ($1s^22s^12p^6$), and thus, L shell is completely filled. The third period (n = 3) starts with sodium and ends at argon in which electrons are successively added into 3s and 3p orbitals. The fourth period (n = 4) starts with potassium in which electrons are added into 4s orbital. In this period, from scandium to zinc the electrons fill up 3d orbital before their 4p orbital gets filled because of energetically favorable process, and hence, these elements are known as 3d transition series. This period ends at krypton with filled 4p orbitals. The fifth period (n = 5) begins with rubidium ($5s^1$) and ends at xenon with filled 5p orbitals. This period also consists of 4d transition series starts with yttrium and ends at cadmium. The sixth period (n = 6) consists of 32 elements: the successive addition of electrons in the order 6s, 4f, 5d, and 6p orbitals occur. This period includes 4f - inner transition series known as lanthanide series starting with cerium to lutetium. Similar to sixth period, in seventh period the electrons enter into 7s, 5f, 6d, and 7p orbitals. This period consists of majority of the artificial radioactive elements. The elements after actinium till Lr are called as actinides. The lanthanides and actinides are located at the bottom of the periodic table due to their similar properties.

10.2 Periodic Properties

10.2.1 Ionization Enthalpy

Ionization enthalpy is defined as the amount of energy needed to remove the valence electron from an isolated atom or a molecule in its ground state. It is expressed in kJ mol^{-1}. The first and second ionization energy of element "X" are the energy required to remove the foremost and second most loosely bound electron from a neutral element and positively charged ion respectively as depicted in the reactions (10.1) and (10.2).

$$X + \text{energy} \longrightarrow X^+ + e^- \quad (10.1)$$

$$X^{+} \longrightarrow X^{2+} + e^{-} \tag{10.2}$$

The ionization energy values are always positive. The second ionization energy is invariably higher than the first ionization energy as it needs more energy to remove the electron from positive ion than from neutral atom. The ionization enthalpy is low for alkali metals and hence exhibits high reactivity. It is high for noble gases because of completely filled stable electronic configuration. Generally, in the period, ionization energy increases from left to right and decreases down in the group. The variation of ionization enthalpy in the periodic table can be correlated with two factors: i) attractive force between the electron and the nucleus, and ii) repulsive force that exist among the electrons.

In case of alkali metals, the outer "ns" electrons are shielded or screened from the nucleus by the inner orbital electrons and thus the valence electrons experience less net positive charge. Thus, the removal of valence electrons requires less energy. In a period, the successive electrons fill up the orbital of the same principal shell and increase in the nuclear charge tend to attract the outermost electrons as we move from left to right. Thus, ionization energy increases. As we move down in group ionization energy decreases because increase in number of shells makes atoms size bigger, and hence, the valence electrons are placed away from the nucleus. In addition, increase in the shielding effect makes the easy removal of electrons.

10.2.2 Electron Affinity

It is described as the amount of energy discharged on the addition of an electron to neutral atom or molecule to form a negative ion. It is depicted in Equation (10.3).

$$X + e^{-} \longrightarrow X^{-} + \text{energy} \tag{10.3}$$

In the periodic table, halogens have high electron affinity enthalpy as they accept one electron to form a stable noble gas electronic configuration. In case of noble gas, electron affinity values are positive because the electron enters the next higher principal shell to make unstable electronic configuration. The electron affinity tends to increase when we move from left to right in a period due to decrease in the atomic size and increase in the effective nuclear charge, whereas it is vice-versa when we move down in a group.

10.2.3 Electro-Negativity

It is known as the tendency of an atom in a molecule to attract the shared pair of electrons towards itself. According to the Pauling scale, the most and least electro-negative elements are fluorine (4.0) and cesium (0.7). Electro-negativity increases across the period due to decrease in atomic size and increase in nuclear charge, whereas electro-negativity decreases down the group due to increase in atomic size.

10.3 Chemical Bonding

Chemical bond holds the atoms and ions in the molecules and crystals. Only the valence electrons are involved in the chemical bond. Majority of the elements in the periodic table obey the octet rule in the chemical bonding which means the atoms tend to have eight electrons in its valence shell to gain the nearest noble gas electronic configuration. Octet rule concludes that the atom is stable when the valence shell is filled with eight electrons. Therefore, the noble gases are chemically inert. The molecules form through different types of primary chemical bonds such as ionic, covalent and metallic bonds, and secondary bonds such as hydrogen bonds and Van der Waals forces.

10.3.1 Ionic Bond

This type of bond forms as a result of transfer of electrons from one atom to another. Because of electro-negativity difference the atoms either gain or lose electrons during this bonding. The electro-static force of attraction between cation and anion makes this bond stronger. Sodium loses electron to chlorine form sodium chloride (NaCl) through ionic bond.

10.3.2 Covalent Bond

The sharing of valence shell electrons between atoms results in covalent bonding. This type of bond typically exists between non-metals. The bonding between two identical atoms is non-polar covalent bond, which exhibits equal sharing of electrons and makes the bond harder to break. Few examples are oxygen (O_2) and nitrogen (N_2) molecules. The bond due to unequal sharing of electrons between atoms is known as polar covalent bond in which the electron spends on the more electro-negative atom. Such type of bond exists in H_2O molecule.

10.3.3 Metallic Bond

It is the chemical bond results due to sharing of electrons between the atoms of metallic elements. This type of bond includes the delocalization of electrons in the crystal lattice. For example, in the case of metallic sodium, the Na^+ ions are surrounded by a pool of electrons.

10.3.4 Hydrogen Bond

It is a bond arises from the dipole-dipole interaction between hydrogen atoms located in between a pair of higher electro-negative atoms. An individual hydrogen bond is weak, whereas many hydrogen bonds together are strong. Hydrogen bond is weaker than ionic or covalent bond but stronger than Van der Waals forces. Example is hydrogen bonding in water.

10.3.5 Van der Waals Forces

It is a weak intermolecular force of attraction in solids, liquids, or gases. This is a distance dependent force and unlike ionic and covalent bond, it does not lead to chemical bond. As the intermolecular distance increases, it disappears.

10.4 Geochemical Classification and Distribution of Elements

There are numerous ways to classify the elements. Among them, the most significant classification of elements is made by Victor Moritz Goldschmidt in 1923 based on the distribution and geochemical aspect of elements to form varieties of compounds [4]. He categorized the elements into lithophile, siderophile, chalcophile, atmophile, and biophile, which are presented in Table 10.1.

10.4.1 Lithophiles

Lithophiles are referred as rock-loving elements because they strongly prefer to make a bond with oxygen and abundantly present in the form of rock-forming silicate minerals. They also concentrated in the mantle. The major and minor lithophiles are listed in the Table 10.1. These elements generally exist in silicates, oxides, phosphates, sulfates, halides, and carbonates.

Table 10.1 Geochemical classification of elements by Goldschmidt [4].

Lithophile	Siderophile	Chalcophile		Atmophile	Biophile
		In meteorites	**Terrestrial**		
Si, Al, O, Ti, Zr, Hf, Th, F, Cl, Br, I, B, Sc, Y, La, Ce, Pr, Nd, Sm, Eu, Gd, Tb, Dy, Ho, Er, Tu, Yb, Lu, Li, Na, K, Mg, Ca, Ba, Sr, Cs, Be, Rb, V, Cr, Mn, Nb, W, Ta, U.	Ni, Co, Fe, P, C, Ru, Rh, Pd, Os, Ir, Pt, Au, Ge, Sn, Mo, Ta.	Major - S, Se, P, As, Cu, Ag, Zn, Cd, V, Cr, Mn, Fe.	Major - S, Se, Te, As, Sb, Bi, Ga, In, Tl, Pb, Zn, Cd, Hg, Cu, Ag, Ni, Pd, Co, Fe, Ru, Mo.	C, H, N, O, Cl, Br, I, He, Ne, Ar, Kr, Xe	Major - C, H, O, N, P, S, Cl, I.
Minor – H, S, P, Sn, Fe, Ga, C, Co, Ni	Minor - As, W, Nb, Se, Te	Minor - Te, Sb, Ti, Ca	Minor - Ge, Sn, Au, Pt, Rh, Ir, Os		Minor - B, Ca, Mg, K, Na, V, Mn, Fe, Cu

10.4.2 Siderophiles

Siderophiles have strong affinity for iron and no affinity for oxygen. The compounds from these elements are formed through metallic bond. They also dissolve in molten iron and found in the inner core of the earth. For example, meteorites include Co and Ni.

10.4.3 Chalcophiles

Chalcophiles exhibit strong affinity for copper and hence known as copper loving elements. They majorly exist in sulfides ores. These elements are also found in selenides, and tellurides. They are concentrated in the core.

10.4.4 Atmophiles

Being highly volatile, they exist in the form of liquid or gas at the earth's surface. They mainly include noble gases and occur in both the atmosphere and hydrosphere.

10.4.5 Biophiles

Biophiles concentrate in living organism. C, H, O, N, P, and S are the major macronutrients.

10.5 Chemical Composition of the Earth

Earth is the only planet which supports life. The atmosphere of earth is comprised of 78% of nitrogen and 21% of oxygen. The earth is categorized into three different layers based on the chemical composition as shown in Figure 10.2.

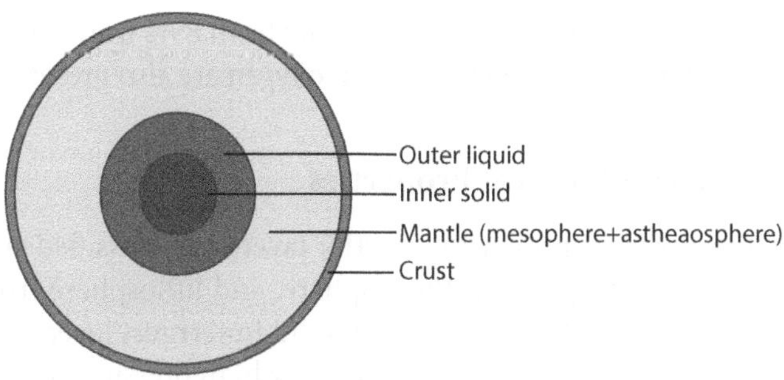

Figure 10.2 Layers of the earth.

10.6 Classification of Earth's Layers

10.6.1 Based on Chemical Composition

Based on the chemical composition, earth's layers are classified into (i) crust, (ii) mantle, and (iii) core. Crust is the thin, easily accessible outermost layer of earth. This forms only 1 % of earth's volume and 0.5% of its mass. The most common elements such as silicon, oxygen, aluminum, and minerals like feldspar are the primary chemicals existing on crust. Silicon dioxide present as the second major chemical. The crust on earth is classified into continental and oceanic crust [5, 6].

i) Continental crust: It is composed of continents. Its thickness and density ranges from 25–70 km and 2.7–3.0 g/cm^3, respectively. This part of the crust mainly consists of granites, feldspar, quartz, biotite, muscovite, and clay minerals.

ii) Oceanic crust: This is composed of oceanic floor. The thickness and density is approximately 5–6 km and 3.0–3.0 g/cm^3, respectively. It consists of silica poor basalt and other ferromagnesian silicate minerals like olivine, pyroxene, and hornblende.

Mantle forms around 84% of the earth's volume. This layer covers from the base of crust to the top layer of core. The zone of the upper mantle is derived from tectonic plates which help to study the earthquake and various parameters such as heat flow and gravity. The upper mantle is majorly comprised of the minerals namelyolivine and pyroxenes. The lower mantle is made up of iron and magnesium silicates like pyroxenes. The lowest layer is known as "D layer" which is rich in aluminum and calcium [5, 7]. The core forms 15% of the earth's volume and 33.5% of its mass. It includes liquid outer layer and solid inner layer. The core is made principally from 89% iron along with 6% nickel. The lighter elements such as sulfur, oxygen are also present [5].

10.6.2 Based on Physical Properties

Based on the physical properties, earth's layers are classified into inner core, outer core, mesosphere, asthenosphere, and lithosphere from inside to outer earth's surface. Inner core is the solid innermost layer of the earth is the inner core region. This exists under very high pressure consisting hot solid iron-nickel mixture in spite of high temperature. Outer core region exists around the inner core with high temperature and pressure in which iron present in the molten state. It exhibits magnetic field.

Mesosphere is the rigid layer below the asthenosphere includes the lower part of mantle. This is also known as intermediate sphere. Here, the rocks are relatively weaker and can undergo deformation easily. In the lower region of mantle, despite of high temperature rocks have comparatively high strength [5, 8, 9]. Asthenosphere is the ductile layer found beneath the lithosphere where the rocks are hotter and convect heat through conduction [5, 8, 9]. Lithosphere is the hard outer layer of the earth composed of crust and the uppermost mantle is often referred as rock layer.

10.7 Spheres of the Earth

Earth is the only unique planet in the solar system containing oxygen and oceans helps in the existence of diverse life on earth. It is a very complex geological zone on which four major distinct spheres are found. They are geosphere or lithosphere, hydrosphere, atmosphere and biosphere. All the four spheres are interconnected.

10.7.1 Geosphere/Lithosphere

The solid outermost portion of the earth constitutes the crust and upper part of the mantle. Lithosphere includes rocks, minerals, and soil. It is classified into oceanic and continental lithosphere. Oceanic crust is associated with the sediments of rocks and sand. The continental lithosphere is made up of continental granitic crust and extended in between 20 and 35 miles. The continental lithosphere is less dense than the oceanic lithosphere. In the outermost lithosphere's layer, the soil rich in oxygen, silicon, and nutrients, whereas the mantle lithosphere consists of oxygen, silicon, iron, and magnesium. The solid crust is made up of rocks which are classified into three different types, namely, (i) metamorphic rocks—originated by applying heat and or pressure on the pre-existing rocks, (ii) igneous rocks—formed from cooling of magma, and (iii) sedimentary rocks—derived from the pre-existing rocks or organic matter by geological processes such as erosion, weathering, dissolution, and precipitation. According to plate tectonics, the continental lithosphere is divided into several major plates. The thermal energy of the mantle lithosphere is responsible for the movement of tectonic plates. Several geological events such as earthquakes, volcanic eruption, and deep ocean trenches are formed by the action of tectonic plates in the lithosphere. Lithosphere influences the temperature difference by interacting with the hydrosphere and atmosphere [10–12].

10.7.2 Hydrosphere

Earth's surface is composed of approximately 71% hydrosphere and 29% land. Earth is known as blue planet because its large part is covered by water. Water is very essential for plants and animals for their survival. Hydrosphere is the most important sphere in which water occur in all the three states. Water exists as lakes, rivers, and oceans on earth. It exists as a well, underground, and aquifers below the ground. In the atmosphere, water vapor is present in the form of fog and clouds. In addition to this, the cryosphere (the frozen part of hydrosphere) is made up of ice bergs, ice caps, and glaciers. In the hydrosphere, water moves continuously through hydrological cycle [13].

10.7.3 Biosphere

It is known as the zone of life on earth because it is composed of all the living things such as plants, animals, and living organisms makes the planet unique. Biosphere is mainly comprised of H, C, O, N, and P elements and is responsible for carbon cycle. The biosphere depends on hydrosphere for water, lithosphere for minerals, and atmosphere for gases. Millions of plants produce oxygen during photosynthesis which in turn used by the other living things for respiration. The food chain and food web are the important aspects where the carnivores and omnivores depend on herbivores for food [14].

10.7.4 Atmosphere

Atmosphere is the gaseous layer surrounding the earth. Earth is the only planet in the solar system which has atmosphere favorable for the existence of life. This layer covers an altitude of 1,600 km from the surface of earth. The atmospheric layer until 60 km from the height of earth's surface has virtually uniform composition. The major species, such as N_2, O_2, and Ar, and the minor species, such as CO_2, He, Ne, CH_4, Kr, H_2, O_3, and Xe, contribute to the atmosphere. Even at increasing distance from the earth, the upper atmosphere is dominated by N_2 and O_2. The earth's atmosphere is divided into different layers based on temperature. They are troposphere, stratosphere, mesosphere, thermosphere, ionosphere, and exosphere [13]. Figure 10.3 shows different spheres of the earth's surface.

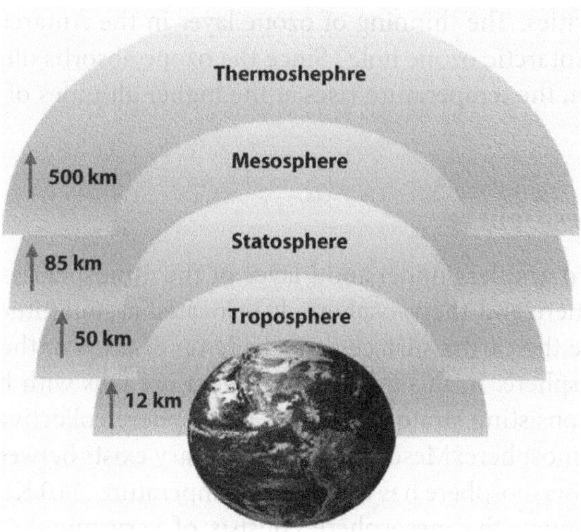

Figure 10.3 Spheres of the earth's surface.

10.7.5 Troposphere

The bottom layer of the atmosphere where we live in is known as troposphere. It exists at an altitude of 12 km from the surface of earth. It is the first largest layer of the earth consisting nearly 90% of the weight of the earth's atmosphere. This portion is constituted by the majority of weather—clouds, snow, and rain. It is comprised of 75% of atmospheric air along with the entire water vapor. The lowest region of troposphere is warmest due to the transfer of heat from the earth's surface, whereas temperature decreases by about 6.5° per kilometer raise in altitude. The boundary exist in the atmosphere that separates the troposphere from stratosphere is known as tropopause. This boundary is expanded to 16 km near the equator and contracted to 8 km at the poles [15, 16].

10.7.6 Stratosphere

Stratosphere is the second layer of the earth's atmosphere as we move upward. It exists at an altitude of 50 km from the tropopause. It is composed of nearly 9.9% of the mass of the earth's atmosphere. Most of the ozone is present in this layer. The ozone layer acts like a shielding layer because it absorbs the harmful UV radiation from the sun thereby protects us from health damage. At polar latitudes, ozone layer is gradually depleting due to the gases released from industries comprising halons and other human activities. The thinning of ozone layer in the Antarctica region is known as "Antarctic ozone hole". Since the ozone absorbs ultraviolet light from the sun, the temperature rises at the higher altitudes of stratosphere [17, 18].

10.7.7 Mesosphere

It is the third and less understood layer of the atmosphere, lies between the stratosphere and thermosphere. It is located at an altitude of around 85 km above the earth's surface. It is made up of 0.1% of the mass of the earth's atmosphere. In this layer, the temperature falls with height. Often the region consisting stratosphere and mesosphere collectively is termed as middle atmosphere. Mesopause is a boundary exists between the mesosphere and thermosphere has the coldest temperature (110 K) in the atmosphere. Although the mesosphere consists of very minute water vapor, highest clouds exist in this sphere at greater heights are called as "noctilucent clouds". In the mesosphere, few meteors exist in the form of pebbles and vaporize to produce heat. The retention of certain constituents of meteors causes the relatively high deposition of iron and other metal atoms in the mesosphere [19, 20].

10.7.8 Thermosphere and Ionosphere

It is the second highest layer of the earth's atmosphere elevated from 85 to 500 km. It is located above the mesosphere and below the exosphere. The lower thermosphere mainly constituted with 80% N_2 and 20% O_2 gases, whereas the upper thermosphere predominantly consists of O_2 gas. The boundary located between the thermosphere and exosphere is known as thermopause. During the strong activity of sun, most of the high energy X-ray and ultraviolet radiations are absorbed and hence the temperature in this sphere reaches as high as 2000 K. The lower portion of the earth is comprised of ionosphere in which the O_2 gas molecules interact with UV radiation leads to the formation of O_2 and O^+. This ionosphere reflects the radio waves. The region comprising thermosphere and exosphere is referred as upper atmosphere [21, 22].

10.7.9 Exosphere

It is the outmost region of the earth's atmosphere exists above the thermosphere and extended in the outer space. The lower part of the exosphere is called as an exobase near which the traces of hydrogen and helium, oxygen, carbon dioxide, and nitrogen are predominant. Since the density of the gas molecules is extremely low, they do not colloid with each other [23].

10.8 Sub-Disciplines of Geochemistry

Modern geochemistry is classified into various sub-disciplines depending on the parameters such as the area of interest, sample type, temperature region, the scientific issues being addressed, etc. Few of them are listed here:

i) Inorganic geochemistry: Inorganic geochemistry provides a vital contribution in analyzing the compositions and structures of geological elements, minerals, soil, and water. The minerals from different types of rocks are mainly composed of silicates. Inorganic geochemistry is one of the most important sub-fields of geochemistry in order to understand the mechanisms of geological processes [24].
ii) Organic and petroleum geochemistry: It includes the study of composition and distribution of organic matter or carbonaceous materials in the lithosphere. Investigation on the evolution of petroleum, fossil fuels, and other hydrocarbons is still the thrust areas of research for organic geochemists [25].
iii) Isotope geochemistry: It deals with the analysis of the absolute and relative abundance of isotopes in the both in the terrestrial and planetary system. The origin of igneous process and rock cycles are interrogated [25].
iv) Cosmo chemistry: It involves the study of distribution of isotopic elements in cosmos [26].
v) Environmental geochemistry: It includes the prediction of both organic and inorganic materials polluting air, ground, and water [27].

10.9 Scope of Geochemistry

Study of geochemistry is one of the thrust and fascinating areas in science. With the knowledge of geochemistry,

- Unknown rocks and minerals can be identified by the quantitative analysis of elements;

- History or origin of the minerals and rocks can be known;
- Accurate compositions of ores, minerals, soil, and rocks are possible to determine;
- Various chemical processes and the reactions in the earth can be understood;
- Crystal structures of various minerals that are existing naturally in the earth can be specified;
- Evolution of the earth and other planets can be learnt;
- Cosmic abundance of the elements can be studied;
- Environmental issues such as greenhouse effect, ozone layer depletion, acid rain, global warming, and various pollutions can be understood and addressed;
- Geological time scale can be quantified;
- Temperature and pressure of magma and geochronology can be analyzed;
- Age of the rocks can be identified from isotopes;
- Ice ages can be investigated.

10.10 Conclusion

The classification of elements based on the atomic weight and geochemical distribution has been understood. The periodic properties such as ionization energy, electro-negativity, and electron affinity influence on the chemical bond formation among the elements. The distribution of elements in different layers present on the earth and beneath the earth is not uniform and depends on the factors like temperature, pressure, and altitude. The sub-fields of the geochemistry have the wide scope to analyze and solve large number of environmental related queries.

References

1. White, W.M., *Geochemistry*, Wiley-Blackwell, Chichester, 2013.
2. White, W.M., History of Geochemistry, in: *Encyclopedia of geochemistry*, pp. 1–15, 2017.
3. Albarede, F., *Geochemistry an introduction*, 2nd Edition, pp. 1–342, Cambridge University Press, New York, 2013.
4. Hollabaugh, C.L., Modification of Goldschmidt's geochemical classification of the elements to include arsenic, mercury, and lead as biophile elements. *Dev. Environ. Sci.*, 5, 9–31, 2007.

5. Fowler, C.M.R., The Earth: Core, mantle and crust, in: *Principles of geologic analysis*, pp. 18–39, 2012.
6. Meissner, R., Introduction: Crust, mantle, lithosphere, and asthenosphere. *Int. Geophys.*, 34, 1–7, 1986.
7. Righter, K. and Drake, M.J., Partition coefficients at high pressure and temperature, in: *Treatise on Geochemistry*, pp. 425–449, 2003.
8. Condie, K.C., The core, in: *Earth as an evolving planetary system*, pp. 135–146, 2016.
9. Pearson, D.G., Canil, D., Shirey, S.B., Mantle samples included in volcanic rocks: xenoliths and diamonds, in: *Treatise on Geochemistry*, pp. 171–275, 2003.
10. Rudnick, R.L., Introduction to Volume 3, in: *Treatise on Geochemistry*, pp. xv–xix, 2003.
11. DiPietro, J.A., An introduction to the geography, geology, and natural history, in formation, collapse, and erosonal decay of mountain systems, in: *Landscape evolution in the United States*, pp. 365–373, 2013.
12. Bina, C.R., Seismological Constraints upon Mantle Composition, in: *Treatise on Geochemistry*, pp. 39–59, 2003.
13. DellaSala, D.A., Freshwater and global change: Wellspring of life, in: *Encyclopedia of the Anthropocene*, pp. 21–24, 2018.
14. Hens, L. and Quynh, L.X., Environmental space, in: *Encyclopedia of ecology*, pp. 1356–1363, 2008.
15. Holton, J.R., Stratosphere/troposphere exchange & structure | Global aspects, in: *Reference Module in Earth Systems and Environmental Sciences Encyclopedia of Atmospheric Sciences*, Second Edition, pp. 257–261, 2015, https://doi.org/10.1016/B978-0-12-382225-3.00394-7.
16. Showman, A.P. and Dowling, T.E., Earth as a planet: Atmosphere and Oceans, in: *Encyclopedia of the solar system*, pp. 423–444, 2014.
17. Brusseau, M.L., Matthias, A.D., Musil, S.A., Bohn, H.L., Physical-chemical characteristics of the atmosphere, in: *Environmental and pollution science*, Third Edition, pp. 47–59, 2019.
18. Lamarque, J.F. and Hess, P., Stratosphere/troposphere exchange & structure | Local processes, in: *Reference Module in Earth Systems and Environmental Sciences, Encyclopedia of Atmospheric Sciences*, Second Edition, pp. 262–268, 2015, https://doi.org/10.1016/B978-0-12-382225-3.00395-9.
19. Varney, R.H. and Kelley, M.C., Polar summer mesopause, in: *Reference module in earth systems and environmental sciences, Encyclopedia of atmospheric sciences*, Second Edition, pp. 436–443, 2015.
20. Liang, J., Chemical composition of the atmosphere of the Earth, in: *Chemical modeling for air resources, Fundamentals, applications, and corroborative analysis*, pp. 3–20, 2013.
21. Solomon, S.C. and Roble, R.G., Thermosphere. Reference module in earth systems and environmental sciences, in: *Encyclopedia of atmospheric sciences*, Second Edition, pp. 402–408, 2015.

22. Hunten, D.M., Venus: Atmosphere, in: *Encyclopedia of the solar system*, Second Edition, pp. 139–148, 2007.
23. Speight, J.G., Chemicals and the Environment, in: *Environmental organic chemistry for engineers*, pp. 1–41, 2017.
24. Trueman, C.N., Rodgers, K.J., McLellan, I.S., Hursthouse, A.S., Geochemistry/Inorganic, in: *Encyclopedia of Analytical Science*, 3rd Edition, pp. 171–181, 2005.
25. Mather, T.A., Geochemistry, in: *Reference module in earth systems and environmental sciences*, pp. 1–3, 2013.
26. Laeter, J.R.D., Cosmochemical applications using mass spectrometry, in: *Encyclopedia of Spectroscopy and Spectrometry*, pp. 359–367, 1999.
27. Abrajano Jr., T.A., Yan, B., Song, J., Bopp, R., Malley, V.O., High-Molecular-Weight Petrogenic and Pyrogenic Hydrocarbonsin Aquatic Environments, in: *Treatise on geochemistry*, vol. 9, pp. 1–50, 2007.

Index

Printed and bound by CPI Group (UK) Ltd, Croydon, CR0 4YY
17/09/2026
14964246-0002